Lecture Notes in Mathematics

Edited by J.-M. Morel, F. Takens and B. Teissier

Editorial Policy
for the publication of monographs

1. Lecture Notes aim to report new developments in all areas of mathematics and their applications- quickly, informally and at a high level. Mathematical texts analysing new developments in modelling and numerical simulation are welcome.

   Monograph manuscripts should be reasonably self-contained and rounded off. Thus they may, and often will, present not only results of the author but also related work by other people. They may be based on specialised lecture courses. Furthermore, the manuscripts should provide sufficient motivation, examples and applications. This clearly distinguishes Lecture Notes from journal articles or technical reports which normally are very concise. Articles intended for a journal but too long to be accepted by most journals, usually do not have this "lecture notes" character. For similar reasons it is unusual for doctoral theses to be accepted for the Lecture Notes series, though habilitation theses may be appropriate.

2. Manuscripts should be submitted (preferably in duplicate) either to one of the series editors or to Springer-Verlag, Heidelberg. In general, manuscripts will be sent out to 2 external referees for evaluation. If a decision cannot yet be reached on the basis of the first 2 reports, further referees may be contacted: The author will be informed of this. A final decision to publish can be made only on the basis of the complete manuscript, however a refereeing process leading to a preliminary decision can be based on a pre-final or incomplete manuscript. The strict minimum amount of material that will be considered should include a detailed outline describing the planned contents of each chapter, a bibliography and several sample chapters.

   Authors should be aware that incomplete or insufficiently close to final manuscripts almost always result in longer refereeing times and nevertheless unclear referees' recommendations, making further refereeing of a final draft necessary. Authors should also be aware that parallel submission of their manuscript to another publisher while under consideration for LNM will in general lead to immediate rejection.

3. Manuscripts should in general be submitted in English. Final manuscripts should contain at least 100 pages of mathematical text and should include
   - a table of contents;
   - an informative introduction, with adequate motivation and perhaps some historical remarks: it should be accessible to a reader not intimately familiar with the topic treated;
   - a subject index: as a rule this is genuinely helpful for the reader.

Continued on inside back-cover

Lecture Notes in Mathematics 1811

Editors:
J.-M. Morel, Cachan
F. Takens, Groningen
B. Teissier, Paris

Springer

Berlin
Heidelberg
New York
Hong Kong
London
Milan
Paris
Tokyo

Jan Stevens

Deformations
of Singularities

Springer

Author

Jan Stevens
Matematik
Göteborgs Universitet
Chalmers Tekniska Högskola
41296 Göteborg, Sweden

e-mail: stevens@math.chalmers.se

Cataloging-in-Publication Data applied for

Bibliographic information published by Die Deutsche Bibliothek

Die Deutsche Bibliothek lists this publication in the Deutsche Nationalbibliografie;
detailed bibliographic data is available in the Internet at http://dnb.ddb.de

Mathematics Subject Classification (2000):
32S05, 32S25, 32S30, 14B05, 14B07, 14B12, 14H20

ISSN 0075-8434
ISBN 3-540-00560-9 Springer-Verlag Berlin Heidelberg New York

Springer-Verlag Berlin Heidelberg New York a member of BertelsmannSpringer
Science + Business Media GmbH

http://www.springer.de

Typesetting: Camera-ready TEX output by the author

SPIN: 10910184   41/3142/du-543210 - Printed on acid-free paper

Preface

Deformation theory has the reputation of being a difficult subject, for which no good literature is available. The main obstacle to understanding most of the existing texts is their level of generality. But many of the problems one has to confront already occur in the well established theories of deformations of compact complex manifolds (Kodaira–Spencer theory) and of universal unfoldings of function germs (Thom–Mather theory). When writing my Habilitationsschrift in Hamburg, of which these notes are an outgrowth, I decided to start with some introductory chapters on deformation theory. The warning to prospective authors of popular books on science, that each mathematical formula will cut down the readership by half, applies mutatis mutandis to the use of cofibred categories. They seem abstract nonsense, but in fact versality has its most natural formulation in these terms. There is a certain discongruity between general theory and practical computations (of which I have been doing quite a lot during the last years). One point I want to make, is that both can be understood as the problem of solving a deformation equation.

Having one's papers (since 1989) on file tempts one to 'recycle' old work. But I do hope that the slow process of revising, which led to the present text, has at least removed some of the mistakes.

It is a pleasure to thank all those who contributed in one form or another to the existence of these notes. Especially I want to thank Kurt Behnke for the many discussions during our joint time in Hamburg. For conversations, which among other things helped me shape my ideas, I am grateful to Duco van Straten, Ragnar Buchweitz, Jan Christophersen, Theo de Jong, Miles Reid and Jonny Wahl (this is a non-exhaustive list). I thank Oswald Riemenschneider and the participants of his 'Seminar über Komplexe Analysis' in Hamburg. I thank the former European Singularity Project and its successor for the environment it created, and its organisers for their efforts. I especially thank Gert-Martin Greuel, Dirk Siersma, and Terry Wall. Thanks to Klaus Altman for finding mistakes in the originally submitted Habilitationsschrift.

I could not have computed so many examples without computer use, more specifically the computer program *Macaulay* [BS]. Therefore thanks to Dave Bayer and Mike Stillman, and David Eisenbud for his scripts. I managed to enlighten the text with some real pictures thanks to the program *surf* by Stephan Endraß [En].

Göteborg, August 2002.

Jan Stevens.

Table of Contents

Introduction

Deformation theory has its origins in the theory of moduli; '[eine Theorie], die für uns, als wir begannen, lange Zeit mit einem Schleier umhüllt war. Riemann sagt, daß bei beliebiger birationaler Transformation des Gebildes nicht nur die Zahl p ungeändert bleibt, sondern (für $p > 1$) auch noch $3p - 3$ Konstanten, die er „Moduln" des Gebildes nennt. Diese Moduln sind einfach die absoluten Invarianten, welche die Normalkurve C_{2p-2} gegenüber linearer Transformationen ihrer homogenen Koordinaten aufweist! ... Es ist doch die lineare Invariantentheorie, die die Probleme beherrscht, aber nur, wenn man sie richtig in Ansatz bringt!' [Kle, p. 310]. The situation for Riemann surfaces is the following: for the topological classification one discrete invariant suffices, the geometric genus p_g, but the analytical structure depends on continuous parameters, called moduli, which one would like to consider as coordinates on a reasonably nice space, whose points are in one-to-one correspondence with the isomorphism classes of Riemann surfaces. KLEIN indicates a way to construct the moduli space as orbit space: it is the quotient of the finite dimensional space parametrising canonically embedded curves (this almost works; to include the hyperelliptic curves one has to use the tricanonical embedding instead). For other classification problems one can try to do the same.

The fundamental discovery of KODAIRA and SPENCER is that already for surfaces moduli spaces in general do not exist [KS]. Instead one has to settle for 'local moduli', as it is sometimes called; in the terminology of these notes, it is the semi-universal deformation of a given object. We start out with a given surface and deform the analytical structure. Now we want a space which not only parametrises all isomorphism classes of nearby structures, but also all germs of continuous families, which contain the given surface as special element. Such a space exists and the 'number of moduli' for the given surface can be defined as the dimension of its tangent space. Nearby surfaces can depend on less moduli. This forces the 'moduli space' to be non-Hausdorff, which means that it cannot exist, at least as the reasonably nice space we wanted. Such exceptional behaviour is the rule in the Thom–Mather theory of unfoldings of functions (cf. [AGV]). Here finite determinacy allows the reduction of the classification problem to a finite dimensional one: the Lie group of k-jets of coordinate transformations acts on k-jets of functions, and

the semi-universal unfolding of a function is obtained from a transversal slice
to the orbit of its jet under this action. For many deformation problems the
picture is in principle the same (cf. [Bi2]), but one has to work with infinite
dimensional spaces, which lead to enormous analytical problems.

A second source of deformation-like problems in algebraic geometry is
the theory of (linear) systems of curves on surfaces and its generalisations,
see [Mu2]. The Zariski tangent space to the family is $H^0(C, N_C)$, where N_C
is the normal bundle, and obstructions lie in $H^1(C, N_C)$; the question, if the
family in question is smooth of dimension $h^0(C, N_C)$, is classically known as
the problem of the completeness of the characteristic linear system. The functorial language to treat such problems has been developed by GROTHENDIECK
[Gro]. He also realised that his formalism is the correct one for deformation
theory in general and that of compact complex manifolds in particular.

The first half of these notes treats the general theory of deformations and
of deformations of singularities in particular. As illustration some quite specific computations are given. The remaining chapters consider more specific
problems, specially on curves and surfaces.

Deformations of singularities (i.e., germs of analytic spaces) can be described in terms of deformations of the local ring, in the sense of [Ge]. The underlying vector space is not changed, but the multiplication map is perturbed.
That the deformed multiplication is again associative is a highly non-trivial
condition. Since the work of SCHLESSINGER (cf. [Art3, Schl2]) and TYURINA
[Ty] we have a direct definition in terms of defining equations, see Chap. 1.
The existence of versal deformations for isolated singularities has been shown
by GRAUERT in a much cited (but not read?) paper [Gra]. Using a power series Ansatz, computations are possible. This is all what can be said in this
generality; starting with a system of equations $(f_1(x), \ldots, f_k(x))$, describing
the singularity, one finds a more complicated system $(F_1(x, t), \ldots, F_k(x, t))$
(where $F_i(x, 0) = f_i(x)$), and in general also equations $g_j(t)$ between the
deformation parameters.

'Grauert löste die zentralen Probleme der Deformationstheorie mit so
schlagender Gewalt, daß der ganzen Theorie darüber fast der Atem ausging.'
[Remmert, Laudatio at the presentation of the von Staudt price to Grauert,
see DMV-Mitteilungen 1/93]. The existence of versal deformations should
really be the beginning of the theory. The situation however is that the existence of a formally versal, formal object is often easy, and the result states
that an analytically versal, analytic object exists. Hard analysis is needed for
the difficult proof, but it sheds no further light on the objects, and in particular it gives no practical way to compute. We give in Chap. 6 BUCHWEITZ'
example of a smooth affine (elliptic) curve, which is formally rigid, but has
non-trivial analytic deformations. In some sense, this is a trivial example,
and it would be interesting to have a similar example with a deformation
problem 'in real life'; of course, in a 'nice world' it does not exist.

It is useful to formulate a deformation problem in terms of a more or less explicit deformation equation; for compact complex manifolds this is achieved by the integrability condition for almost complex structures, see e.g. [Ku]:

$$\bar{\partial}\vartheta + \tfrac{1}{2}[\vartheta,\vartheta] = 0.$$

The most practical way to obtain concrete solutions, is to use a power series Ansatz. As one wants an answer in finite time, one really looks for polynomial solutions. The equation can also be used to obtain existence results, say by using some form of the implicit function theorem. For deformations of singularities one can get the same type of deformation equation from the general theory of the cotangent complex. In concrete computations it boils down to the formalism of lifting relations.

Knowing the existence of versal deformations of singularities we can go on and ask questions about the structure of the base space (is it reduced, what is the number of components?), or ask if for some t_0 the fibre $(F_1(x,t_0)$, ..., $F_k(x,t_0))$ is smooth. In general these question cannot be answered, because the equations are just one enormous mess. However, for specific, well-chosen examples things look better; in most cases the distinguishing factor is some extra symmetry on the equations, which extend through the whole computation. In hand calculations the symmetry can be used to organise them; on a machine the symmetry enters mostly indirectly, in that only certain monomials can occur in an expression, which is therefore smaller and easier to understand.

A striking example of the importance of visible symmetry, is the contrast between the ease, with which in Chap. 3 the versal deformation of L_n^n (the curve, consisting of the coordinate axes in $\mathbb{C}^n$) is computed, and long computations in Chap. 11 for the isomorphic curve in determinantal form (only for $n = 4$; these computations are a prelude to those for the monomial curve (t^4, t^5, t^6, t^7), which is in a natural way determinantal).

It is desirable to have a systematic way to write down equations. For the famous example of the cone over the rational normal curve of degree four, there are two determinantal representations, and in fact they lead to the two components of the base space. More examples are given in Chap. 12 on *formats*. Presumably, there is no general statement, and we are spoilt by the simple examples that are found first.

Explaining smoothing components (of rational surface singularities) is the goal of KOLLÁR's conjectures [Kol]. Originally it was thought that (if true) they would give a method to find the smoothing components, but apart from the case of quotient singularities [KSh], it seems that one has first the components, and then the corresponding P-modifications. The pessimistic view is that one impossible problem is replaced by another. The interest of the conjectures is the new understanding, to which they lead. In Chap. 14 I give a number of examples of non rational singularities with P-modifications, explaining the components, and extend the conjectures to all surface singularities.

For curve singularities no interpretation of smoothing components is known. In this case all smoothing components have the same dimension. Chap. 13 contains the first example of a curve with several smoothing components: it is L^6_{14}, 14 lines in general position through the origin in $\mathbb{C}^6$. I originally performed the calculation to decide whether the singularity was smoothable at all. The case of L^6_{14} was left open in my earlier work on smoothability of certain cones over points [St1].

Cones over curves form the subject of the last two chapters. Powerful methods exist to compute T^1 for surface singularities, without using explicit equations. For cones over curves the bundle of principal parts comes in, and with it WAHL's Gaussian map (cf. [Wa6]). The computation of $T^1(-1)$ is the most difficult; much of the work on the Gaussian map is connected with vanishing results for this case. The most complete results on interesting deformations are obtained by Sonny TENDIAN [Te1]. With a trick, which basically is contained in [Mu3], one sees that for non hyperelliptic curves, embedded with a non-special line bundle L, the dimension of $T^1(-1)$ of the cone equals $h^0(C, N_K \otimes L^{-1})$, where N_K is the normal bundle of C in its canonical embedding. For low genus this gives quite precise information, because then the normal bundle N_K is easy to describe.

If S is a surface with C as hyperplane section, then one can degenerate S to the projective cone over C, or from another point of view, deform the projective cone over C to S; PINKHAM calls this construction 'sweeping out the cone' [Pin1]. Surfaces with hyperelliptic hyperplane sections were already classified by CASTELNUOVO, and the supernormal surfaces among them have degree $4g + 4$ [Cas]. They are rational ruled surfaces, and such surfaces come in two deformation types; therefore there are at least two smoothing components. A computer computation of the versal deformation in negative degree with *Macaulay* [BS] gave for an example with $g = 2$ the number of 32 smoothing components. I show that cones over hyperelliptic curves of degree $4g + 4$ have 2^{2g+1} smoothing components (the case $g = 3$ is exceptional).

1 Deformations of singularities

The definition of a deformation involves the notion of flatness, which accounts for the difficulties in explaining and understanding it. In the mid sixties MUMFORD wrote: "The concept of flatness is a riddle that comes out of algebra, but which is technically the answer to many prayers."[Mul, p. 295]. Intuitively, in a flat family the fibres depend continuously on the points of the parametrising base space. We assume that the reader is familiar with flat morphisms [Ha, III.9], [Fi, 3.11]; the purpose of this section is to show the relevance for deformation theory. After some examples we eventually define a deformation of a space (germ) X_0 as a flat map $\pi\colon X \to S$ with $X_0 \cong \pi^{-1}(0)$. The term *deformation* is a convenient way of speaking, which emphasises the special role of X_0; if we consider the general fibre X_s, $s \neq 0$, as primary object, we speak of a *degeneration*, or *specialisation*.

Example. Consider a quartic curve C_0 in $\mathbb{P}^3$ with a double point. To be specific, let C_0 be the image of the map $f\colon \mathbb{P}^1 \to \mathbb{P}^3$ given by

$$(x_0 : x_1 : x_2 : x_3) = (s_1^4 - s_2^4 : s_1^3 s_2 : s_1^2 s_2^2 : s_1 s_2^3) .$$

The curve C_0 is a complete intersection with equations

$$x_2^2 - x_1 x_3 , \qquad x_1^2 - x_0 x_2 - x_3^2 .$$

The double point lies in $(1 : 0 : 0 : 0)$.

By perturbing the map f to f_t we obtain in general a smooth rational quartic curve C_t. To see what happens in a neighbourhood of the double point, we take local coordinates (x, y, z), and equations $y = 0$, $xz = 0$. The curve germ $(C_0, 0)$ is the image of the multigerm $f\colon \mathbb{C} \cup \mathbb{C} \to \mathbb{C}^3$, given by $f(s_1) = (s_1, 0, 0)$ and $f(s_2) = (0, 0, s_2)$. Now consider the map $F\colon (\mathbb{C} \cup \mathbb{C}) \times \mathbb{C} \to \mathbb{C}^3 \times \mathbb{C}$, defined as $F(s_1, t) = (s_1, 0, 0, t)$ and $F(s_2, t) = (0, t, s_2, t)$. For $t \neq 0$ we have two skew lines in $\mathbb{C}^3 \times \{t\}$, which can be described by four equations:

$$yx = 0 , \qquad zx = 0 , \qquad y(y - t) = 0 , \qquad z(y - t) = 0 .$$

If we put $t = 0$ in these equations, we do not get the equations $y = 0$, $xz = 0$ of the image of f, but four equations $yx = zx = y^2 = zy = 0$; they describe the same plane curve, but with an embedded point sticking out in the y-direction. We remark that the image X of F in $\mathbb{C}^3 \times \mathbb{C} = \mathbb{C}^4$ consists of two planes with one point in common, the simplest example of a non-normal isolated surface singularity.

The family C_t is arguably not a family at all, as it cannot be described by a morphism $\pi\colon X \to \mathbb{C}$; however such families were quite common in classical Italian geometry, and HARTSHORNE gives a complicated definition of an *algebraic family of varieties* [Ha, p. 263] to cover just this type of family. Furthermore, in deforming or unfolding map (multi)-germs, it is quite natural to pull two lines apart, but it does not give a deformation of the image.

The complete intersection C_0 also belongs to the family of all complete intersections of quadrics in $\mathbb{P}^3$, which is a flat family. The general curve of this family is a curve of type $(2,2)$ on a smooth quadric (an elliptic curve). In this context we recall that a family $\pi\colon X \to T$ of projective curves over an integral scheme T is flat if and only if the degree and arithmetic genus of X_t are independent of $t \in T$ [Ha, III.9.9]; for curves these two numbers determine the Hilbert polynomial.

Classically, the double point on the rational quartic C_0 was called an *improper node* with respect to the family of rational curves, and a *proper node* w.r.t. the family of curves of type $(2,2)$ on a quadric. SEVERI distinguishes in [Sev, Anhang G] between the two families by looking at limiting positions of secants.

We define a family with special fibre X_0 over a base space S (containing a point 0) to be a holomorphic map (or germ) $\pi\colon X \to S$, such that X_0 is isomorphic to $\pi^{-1}(0)$. In particular, in the local situation this means that we have equations $F_1(x,s), \ldots, F_k(x,s)$, whose restrictions $F_1(x,0), \ldots, F_k(x,0)$ generate an ideal, isomorphic to the ideal of X_0. This type of family is in general still not nice enough to define deformations.

Example. Let X_0 be the singularity L_3^3 consisting of the coordinate axes in $\mathbb{C}^3$; it is zero locus of the functions $f = xy$, $g = xz$ and $h = yz$. Consider the one parameter family given by

$$F = xy - t^2, \qquad G = xz - t^2, \qquad H = yz - t^2.$$

For $t \neq 0$ the space X_t consists only of the two points (t,t,t) and $(-t,-t,-t)$; we see an abrupt change in the dimension of the fibre. If we take the closure in $\mathbb{P}^3$ we find three points more with multiplicity 2 at infinity — the intersection of three quadrics has degree 8 by Bézout's Theorem.

For $t = 0$ we have more equations than the codimension of X_0. Although the three equations f, g and h are linearly independent, there exists non trivial relations: we have $f \cdot z - g \cdot y = 0$, and $f \cdot z - h \cdot x = 0$, and these relations generate the $\mathbb{C}\{x,y,z\}$-module of all relations. If we try to extend the relations to include the variable t, we find $F \cdot z - G \cdot y = t^2(z - y)$, and $F \cdot z - H \cdot x = t^2(z - x)$. For $t \neq 0$ we can divide by t^2 and take new generators of the ideal; doing this in the family means cutting away the fibres over zero: the ideal $(z - y, z - x, x^2 - t^2)$ describes the lines (t, t, t) and $(-t, -t, -t)$, which are a finite cover of the t-axis. We cannot lift the relations.

Now consider the family

$$F = xy , \qquad G = xz , \qquad H = yz + ty + tz .$$

For $t \neq 0$ the space X_t consists of the x-axis and the smooth hyperbola $(y + t)(z + t) - t^2$, passing through the origin. In this case the projective closure is a (flat) family of (reducible) cubic rational curves. Now we can lift the relations:

$$F \cdot z - G \cdot y = 0 , \qquad F \cdot (z + t) + G \cdot t - H \cdot x = 0 .$$

Example: the cone over the rational normal curve. Let C_d be the rational normal curve of degree d in $\mathbb{P}^d$: take homogeneous coordinates $(s{:}t)$ on $\mathbb{P}^1$, and embed $\mathbb{P}^1$ in $\mathbb{P}^d$ via $z_i = s^{d-i}t^i$, $i = 0, \ldots, d$. Considered as map between affine varieties, this formula realises the affine cone X_d over C_d as cyclic quotient of $\mathbb{C}^2$ under the action $\frac{1}{d}(1, 1)$. This notation means that $\mathbb{Z}/d\mathbb{Z}$ acts as $(s, t) \mapsto (\varepsilon s, \varepsilon t)$ with ε a primitive d-th root of unity. The monomials $s^{d-i}t^i$ generate the ring of invariants of the induced action on $\mathbb{C}[s, t]$ or $\mathbb{C}\{s, t\}$.

Obviously on X_d one has

$$\mathrm{Rank} \begin{pmatrix} z_0 & z_1 & \cdots & z_{d-1} \\ z_1 & z_2 & \cdots & z_d \end{pmatrix} \leq 1 .$$

An easy combinatorial argument shows that the 2×2 minors $f_{ij} = z_i z_j - z_{i+1} z_{j-1}$ generate the ideal of X_d. The numbers of generators is $\binom{d}{2}$, whereas the codimension of X_d is only $d - 1$.

To obtain relations between these equations (a.k.a. *syzygies*) we look at the matrix

$$\begin{pmatrix} z_0 & z_1 & \cdots & z_{d-1} \\ z_0 & z_1 & \cdots & z_{d-1} \\ z_1 & z_2 & \cdots & z_d \end{pmatrix} .$$

The maximal minors vanish identically; on the other hand, row expansion of a minor yields a linear combination of equations. As we can also double the second row, we get $2\binom{d}{3}$ relations, which in fact generate the module of relations. The first three columns of the matrix above give the syzygy $f_{13}z_0 - f_{03}z_1 + f_{02}z_2 = 0$. If $z_1 \neq 0$, we can express f_{03} as combination of f_{02} and f_{13}: $f_{03} = (z_0/z_1)f_{13} + (z_2/z_1)f_{02}$. In a similar way all equations are expressible in the $d - 1$ equations $f_{i-1,i+1}$, $i = 1, \ldots, d - 1$, but only outside the coordinate hyperplanes. Under deformation this property should be preserved. As the coefficients like z_0/z_1 may also vary, it is better to work directly with the syzygies. So the deformation of the other equations should be determined by the way the $f_{i-1,i+1}$ are deformed; this in turn puts conditions on the possible deformations of the $f_{i-1,i+1}$. To understand this we look at the space Z_d defined by these equations; it has two irreducible components: firstly of course X_d, and furthermore a second component Y_d, the (x_0, x_d)-plane with a multiple structure: if $x_i \neq 0$ for some $0 < i < d$, then the equation $f_{i-1,i+1}$ gives that $x_{i-1} \neq 0$ and $x_{i+1} \neq 0$ also, so by induction all coordinates are non-zero and we have a point on X_d. The $\binom{d-1}{2}$ extra equations are needed to cut away Y_d. If we deform the equations $f_{i-1,i+1}$ arbitrarily, then we are likely to get a smooth and therefore irreducible space. A necessary condition to get in this way a deformation of X_d is that the fibres have (at least) two components: in $Z_d \to S$ we have $Z_d = X_d \cup Y_d$.

Let $\pi\colon (X,0) \to (S,0)$ be a map of analytic germs, and suppose an embedding of the fibre $(X_0,0) = (\pi^{-1}(0),0) \subset (\mathbb{C}^N,0)$ is given. Then π can be realised as the composition of an embedding $(X,0) \subset (S \times \mathbb{C}^N,0)$ and the projection of $(S \times \mathbb{C}^N,0)$ onto the first factor (for a proof see [Fi, 0.35]). Fix such a representation. Let $(F_1, \ldots, F_k)$ generate the ideal I of $(X,0)$ in $P := \mathcal{O}_{S,0} \widehat{\otimes} \mathbb{C}\{z_1, \ldots, z_N\}$. The $f_i(z) := F_i(0,z)$ generate the ideal I_0 of $(X_0,0)$ in $P_0 := \mathbb{C}\{z_1, \ldots, z_N\}$.

Definition. The map $\pi\colon (X,0) \to (S,0)$ is flat in 0, if every relation $\sum f_i r_i = 0$ between the f_i lifts to a relation $\sum F_i R_i = 0 \in P$ between the F_i. A map $\pi\colon X \to S$ between complex spaces is flat, if it is flat in every point of X.

To connect this definition with the usual algebraic notion of flatness, recall that a module M over a ring R (commutative, with unit) is called flat, if for every short exact sequence of R-modules

$$0 \longleftarrow N'' \longleftarrow N \longleftarrow N' \longleftarrow 0$$

the induced sequence

$$0 \longleftarrow N'' \otimes_R M \longleftarrow N \otimes_R M \longleftarrow N' \otimes_R M \longleftarrow 0$$

is again exact, or in other words, the functor $- \otimes_R M$ on the category of R-modules is exact. An equivalent condition is that $\mathrm{Tor}_1^R(M,N) = 0$ for every R-module N. If we have a local homomorphisms $R \to R'$ of Noetherian local rings flatness of a finite R'-module M is already characterised by the property $\mathrm{Tor}_1^R(M, R/\mathfrak{m}_R) = 0$. A proof can be found in [Hi2]: for Artinian modules one uses induction on the length, in the general case on the dimension of the support.

Proposition. *The map $\pi\colon (X,0) \to (S,0)$ is flat at 0 (according to the definition above) if and only if $\mathcal{O}_{X,0}$ is a flat $\mathcal{O}_{S,0}$-module.*

Proof. To compute $\mathrm{Tor}_1^S(\mathcal{O}_{X,0}, \mathbb{C})$, where Tor^S stands for $\mathrm{Tor}^{\mathcal{O}_{S,0}}$, we look at a free $\mathcal{O}_{S,0}$-resolution of $\mathcal{O}_{X,0}$:

$$0 \longleftarrow \mathcal{O}_{X,0} \longleftarrow P \xleftarrow{F} P^k \xleftarrow{R} P^l \longleftarrow \cdots,$$

where F is the vector $(F_1, \ldots, F_k)$, and the columns of R generate the module of relations. We tensor this exact sequence with $\mathbb{C}$ to get a complex

$$0 \longleftarrow \mathcal{O}_{X_0,0} \longleftarrow P_0 \xleftarrow{f} P_0^k \xleftarrow{R(0)} P_0^l \longleftarrow \cdots .$$

The vanishing of $\mathrm{Tor}_1^S(\mathcal{O}_{X,0}, \mathbb{C})$ is equivalent to exactness of this complex at P_0^k. Therefore we ask if the columns of $R(0)$ generate the module of relations between the f_i. If they do, then the columns of R are lifts of generators, so every relation lifts; conversely, if every relation lifts, it lifts to an element of the module generated by the columns of R, and by restriction the columns of $R(0)$ generate. $\qquad\square$

Let I be the ideal, generated by F in P. Tensoring the exact sequence $0 \leftarrow \mathcal{O}_{X,0} \leftarrow P \leftarrow I \leftarrow 0$ with $\mathbb{C}$ gives rise to the long exact sequence

$$\mathrm{Tor}_1^S(\mathcal{O}_{X,0}, \mathbb{C}) \quad \longleftarrow \quad \mathrm{Tor}_1^S(P, \mathbb{C}) \quad \longleftarrow \quad \mathrm{Tor}_1^S(I, \mathbb{C}) \quad \longleftarrow \quad \mathrm{Tor}_2^S(\mathcal{O}_{X,0}, \mathbb{C}) \ ,$$
$$\| \qquad\qquad\qquad\qquad \| \qquad\qquad\qquad\qquad\qquad\qquad \|$$
$$0 \qquad\qquad\qquad\qquad 0 \qquad\qquad\qquad\qquad\qquad\qquad 0$$

so I is a flat $\mathcal{O}_{S,0}$-module. By induction we obtain:

Lemma. *Every free resolution of $\mathcal{O}_{X_0,0}$ lifts to a free resolution of $\mathcal{O}_{X,0}$.*

We note some properties of flat maps [Fi, 3.13–3.21].

1) Flatness is preserved under base change: consider a Cartesian square of complex spaces and holomorphic maps:

$$\begin{array}{ccc} X' & \xrightarrow{f} & X \\ \downarrow{\scriptstyle \pi'} & & \downarrow{\scriptstyle \pi} \\ S' & \xrightarrow{g} & S \end{array}$$

Take a $p' \in X'$ and let $f(p') = p$. If π is flat in p, then π' is flat in p'.

2) If $\pi: X \to S$ is flat, then for every $p \in X$ the dimension formula:

$$\dim_p X = \dim_{\pi(p)} S + \dim_p X_0$$

holds.

3) Every flat holomorphic map is open.

4) Let $\pi: X \to \mathbb{C}^k$ be an open holomorphic map. If π is reduced in $p \in X$, then X is reduced in p and π is flat in p. If $\mathcal{O}_{X,p}$ is Cohen-Macaulay, then π is flat in p.

5) A finite holomorphic map $\phi: X \to Y$ is flat if and and only if $\phi_* \mathcal{O}_X$ is a locally free sheaf.

Definition. A *deformation* of a germ $(X_0, 0)$ is a flat map-germ $\pi: (X, 0) \to (S, 0)$, such that $(X_0, 0)$ is isomorphic to the fibre $(\pi^{-1}(0), 0)$ under a given isomorphism $i: (X_0, 0) \to (\pi^{-1}(0), 0)$.

Remark. We used before $(X_0, 0)$ as notation for the fibre over 0 of $\pi: (X, 0) \to$ $(S, 0)$; the definition above only involves an embedding $i: (X_0, 0) \to (X, 0)$, with $(\pi^{-1}(0), 0)$ as image. We may use the map i to identify $(X_0, 0)$ with $(\pi^{-1}(0), 0)$, but it can be useful not to do this, when considering different deformations of the same space.

Definition. A morphism between two deformations $\pi: (X, 0) \to (S, 0)$ and $\pi': (X', 0) \to (S, 0)$ of $(X_0, 0)$ over the same base $(S, 0)$ is a morphism $f: (X, 0) \to (X', 0)$ over $(S, 0)$ (which means that $\pi' \circ h = \pi$), compatible with the embeddings $i: (X_0, 0) \to (X, 0)$ and $i': (X_0, 0) \to (X', 0)$, i.e., $h \circ i = i'$.

The conditions imply that f is an isomorphism; this follows from the following lemma on flatness.

Lemma. *Let $A \to B$ be a local homomorphism of Noetherian local rings, let $u: N \to N'$ be a homomorphism of finitely generated B-modules. Let $k = A/\mathfrak{m}_A$. If $\bar{u}: N \otimes_A k \to N' \otimes_A k$ is bijective, and N' is A-flat, then u is an isomorphism.*

Proof. Let $K' = \operatorname{coker} u$, and tensor the exact sequence

$$N \longrightarrow N' \longrightarrow K' \longrightarrow 0$$

with k. Then $K' \otimes_A k = 0$, so $K' = \mathfrak{m}_A K'$, and therefore $K' \subset \cap_\nu \mathfrak{m}_A^\nu K' = 0$. Put $K = \ker u$. As $\operatorname{Tor}_1^A(N', k) = 0$, the sequence

$$0 \longrightarrow K \otimes_A k \longrightarrow N \otimes_A k \longrightarrow N' \otimes_A k \longrightarrow 0$$

is exact. Hence $K = 0$. □

Definition. Let $\pi: (X, 0) \to (S, 0)$ be a deformation of $(X_0, 0)$, and let $f: (T, 0) \to (S, 0)$ be a holomorphic map. The *induced deformation* is the flat map $f^*(\pi): (X \times_S T, 0) \to (T, 0)$.

Definition. A deformation $\pi: (X, 0) \to (S, 0)$ of $(X_0, 0)$ is called *miniversal* or *semi-universal* if every deformation $\rho: (Y, 0) \to (T, 0)$ of $(X_0, 0)$ is isomorphic to a deformation $f^*(\pi)$, for some map $f: (T, 0) \to (S, 0)$; the map f may not be unique, but it is required that its derivative df is unique.

Definition. A singularity $(X_0, 0)$ is *rigid*, if every deformation $\pi: (X, 0) \to (S, 0)$ of $(X_0, 0)$ is isomorphic to the trivial deformation $p_2: (X_0 \times S, 0) \to (S, 0)$.

Example (continued): the cone over the rational normal curve. As we have seen, the equations for X_d are

$$\operatorname{Rank} \begin{pmatrix} z_0 & z_1 & \cdots & z_{d-1} \\ z_1 & z_2 & \cdots & z_d \end{pmatrix} \leq 1,$$

while the relations are obtained by doubling a row and taking 3×3 minors. Because we have a recipe to obtain the relations from the entries of the matrix, *every deformation of the entries of the matrix defines a deformation of X_d.*

The most general way to deform the matrix for X_d (modulo coordinate transformations) is as follows:

$$\text{Rank} \begin{pmatrix} z_0 & z_1 + t_1 & \cdots & z_{d-1} + t_{d-1} \\ z_1 & z_2 & \cdots & z_d \end{pmatrix} \leq 1 \,.$$

The total space of this deformation is via an obvious coordinate transformation isomorphic to the *generic determinantal*

$$\text{Rank} \begin{pmatrix} x_1 & x_2 & \cdots & x_d \\ y_1 & y_2 & \cdots & y_d \end{pmatrix} \leq 1 \,.$$

We remark that the generic determinantal is an example of a rigid singularity. It is obvious that all deformations of the matrix are trivial in this case, but one has to show that there are no other deformations.

The cone X_d has other deformations. For $d = 4$ we can also write the six equations as the 2×2 minors of a symmetric 3×3 determinantal, and we obtain a deformation

$$\text{Rank} \begin{pmatrix} z_0 & z_1 & z_2 \\ z_1 & z_2 + s_2 & z_3 \\ z_2 & z_3 & z_4 \end{pmatrix} \leq 1 \,.$$

The total space is the cone over the Veronese embedding of $\mathbb{P}^2$ in $\mathbb{P}^5$. This is PINKHAM's famous example; the miniversal deformation consists of the two components just described [Pin1]. The base space has embedding dimension 4, and is given by the three equations $s_2 t_1 = s_2 t_2 = s_2 t_3 = 0$: a line intersecting a three-dimensional linear subspace transversally.

For $d > 4$ the base space consists of the determinantal component and an embedded component at the origin.

Example: hypersurface singularities. Let $f : (\mathbb{C}^{n+1}, 0) \to (\mathbb{C}, 0)$ have a singularity at the origin, and let $(X, 0) = (f^{-1}(0), 0)$. The function f generates the ideal I of $(X, 0)$, and as there are no relations between generators, the flatness condition is empty: every deformation of the defining equation gives a deformation of the space $(X, 0)$.

The notion of isomorphism is the same as that of contact equivalence in the theory of unfolding of functions, for which [AGV] is a good reference. Suppose the singularity is isolated. If $g_1, \ldots, g_\tau$ is a basis of the $\mathbb{C}$-vector space

$$T_X^1 = \mathcal{O}_{n+1} / \left(f, \tfrac{\partial f}{\partial x_0}, \ldots, \tfrac{\partial f}{\partial x_n} \right) ,$$

then $F(x, t) = f + t_1 g_1 + \cdots + t_\tau g_\tau$ defines the miniversal deformation of $(X, 0)$: we have $(\mathcal{X}, 0) = (F^{-1}(0), 0) \subset (\mathbb{C}^{n+1} \times \mathbb{C}^\tau, 0)$ and the map $\pi : (\mathcal{X}, 0) \to (\mathbb{C}^\tau, 0)$ is induced by the second projection.

More generally, if $f\colon(\mathbb{C}^{n+k},0)\to(\mathbb{C}^k,0)$ defines a complete intersection $(X,0)$, then the *Koszul complex* on the components f_i of f resolves the ring $\mathcal{O}_{X,0}$. In particular, the relations between the f_i are generated by the obvious ones: $f_if_j - f_jf_i = 0$, or in vector notation:

$$(f_1,\ldots,f_i,\ldots,f_j,\ldots,f_k)\cdot(0,\ldots,f_j,\ldots,-f_i,\ldots,0)^t = 0\,.$$

The proof of this statement is easy: if $\sum f_i r_i = 0$ is a relation, we can restrict to $f_1 = \ldots = f_{k-1} = 0$, and find $f_k r_k \equiv 0 \bmod (f_1, \ldots, f_{k-1})$. As f_k is not a zero divisor in $\mathcal{O}_{n+k}/(f_1, \ldots, f_{k-1})$, we have $r_k \in (f_1, \ldots, f_{k-1})$, and subtracting suitable multiples of Koszul relations from $\sum f_i r_i$ gives a relation not involving f_k; by induction the result follows.

For any deformation $F(x,t)$ of $f(x)$ we can lift the relation $f_if_j - f_jf_i = 0$ to the relation $F_iF_j - F_jF_i = 0$, and therefore the flatness condition is always satisfied: every deformation of the map f defines a deformation of the complete intersection singularity $(X,0)$.

Example: codimension two Cohen-Macaulay singularities. In this case one has a structure theorem for the finite free resolution (cf. the Introduction of [BE]), which is based on the following theorem of Hilbert–Burch:

Theorem. *Let R be a commutative Noetherian ring, $I \subset R$ an ideal with free resolution*

$$0 \longleftarrow R/I \longleftarrow R \overset{f}{\longleftarrow} R^{t+1} \overset{r}{\longleftarrow} R^t \longleftarrow 0\,.$$

Then the ideal I is a multiple of the ideal, generated by the $t \times t$-minors of the matrix r: there exists a non zero divisor $a \in R$ such that the i-th entry f_i of f equals $a\Delta_i$, where Δ_i is the minor obtained by deleting the i-th row.

In particular, if the codimension is exactly two, then the non zero divisor a is a unit, and the Δ_i generate the ideal I. If $R = \mathbb{C}\{x\}$ and R/I is a Cohen-Macaulay algebra, it has a resolution of the given form. By the theorem a deformation of R/I over a basis S can be described by a deformed matrix r_S with entries in $R\widehat{\otimes}\mathcal{O}_S$.

Example: hyperplane sections. Let $(X,0)$ be an arbitrary germ and $f \in \mathfrak{m}$ a non-zero divisor. In this situation the map $f\colon(X,0)\to(\mathbb{C},0)$ is flat, so f defines a one-parameter deformation of its fibre over the origin. The case of two planes in $\mathbb{C}^4$ meeting in one point, as in the first example, shows that the analytic structure of the fibre may differ from what one naively expects.

The term hyperplane section really refers to a specific embedding of X in some $\mathbb{C}^N$. As we can use the function f as a coordinate in an embedding, we can also in the abstract setting call the fibre $f^{-1}(0)$ a hyperplane section. In general it will be a special hyperplane section. The notion of general hyperplane section $h\colon(X,0)\to(\mathbb{C},0)$ can be made intrinsic (following REID [Re1, 2.5]) by requiring that h projects onto a general element of $\mathfrak{m}/\mathfrak{m}^2$.

Example: cones. Let X_0 be the affine cone over a projective variety $V_0 \subset \mathbb{P}^N$. As equations for X_0 we can take homogeneous polynomials. In particular X_0 admits a good $\mathbb{C}^*$-action, so the general theory of [Pin1] applies, and the deformations are also graded. We say that a (quasi-homogeneous) deformation has *negative weight* if the equations are perturbed with terms of lower degree; the deformation parameters can be given positive degrees. The way to remember this is that an equation $F(x, t)$ is in first order approximation $f(x) + tf'(x)$ with $f'(x) = \frac{\partial F}{\partial t}|_{t=0}$, so the degree of the perturbation is lowered by the degree of t. In a deformation of negative weight of a cone the hyperplane section at infinity is not changed.

Consider the plane curve D_4 with its miniversal deformation

$$x^3 + y^3 + b_1 xy + a_2 x + c_2 y + b_3 \; ,$$

where the lower index of the deformation variables denotes their degree. By specialising to a $\mathbb{C}^*$-orbit in the base space like $(b_1, a_2, c_2, b_3) = (s, 0, 0, s^3)$ we get a quasi-homogeneous 1-parameter deformation, whose equation can also be read as the homogeneous equation of a plane curve with as hyperplane section the three points over which D_4 is a cone. Conversely the plane curve $x^3 + y^3 + sxy + s^3$ gives us an affine cone which is the total space of a quasi-homogeneous 1-parameter deformation.

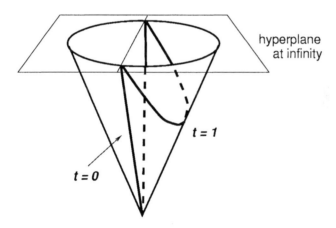

Fig. 1.1. Sweeping out the cone

We have seen an example of a general mechanism which is especially useful if the deformation involved has weight -1, so the deformation parameter has degree 1 and the equations are again homogeneous. The total space is the cone over a projective variety V of one dimension more with V_0 as hyperplane section. Conversely given a projective variety V with V_0 as hyperplane section we can consider the equations as affine equations with t as deformation parameter. The affine planes $\{t = \text{const}\}$ intersect at infinity. A different way to look at it is to consider the projective cone over V; the parallel hyperplanes become a pencil whose base cuts out V_0 on V. The special plane $t = 0$ passes through the vertex of the cone, while the others do not. Projection from the vertex shows that the pair $(C(V) \cap \{t = \text{const}\}, V_0)$ is isomorphic to the pair (V, V_0). PINKHAM calls this construction *sweeping out the cone* [Pin1, 7.6.iii].

2 Standard bases

The construction of Gröbner bases is fundamental for computer computations in algebraic geometry. The basic algorithm is due to BUCHBERGER [Buchb], who introduced the term Gröbner basis. It can be understood as computing the specialisation of a projective variety to one, defined by much simpler equations. The analog in the local case is the specialisation to the tangent cone, or more generally a weighted tangent cone. In this context HIRONAKA introduced the term standard basis [Hi1].

The basic definitions are the almost same in both cases, but in some sense opposite. In the power series case we are interested in terms of lowest order, while in the global case the highest order part is important. We start with the latter one.

Let S be the polynomial ring $\mathbb{C}[x_1, \ldots, x_n]$. The monomials x^a form a $\mathbb{C}$-basis of S: each $f \in S$ can be written as sum $f = \sum_{a \in \mathbb{N}^n} f_a x^a$, with $f_a \in \mathbb{C}$. Let $l : \mathbb{N}^n \to \mathbb{R}_+$ be an injective linear form, $l(a) = l_1 a_1 + \cdots + l_n a_n$. With it we define a total order $<_l$ on $\mathbb{N}^n$ and therefore also on the set of monomials:

$$x^a <_l x^b \iff l(a) < l(b) .$$

This order can be extended to free S-modules, by considering the generators as new variables: if $M = Se_1 \oplus \cdots \oplus Se_m$, then $x^a e_j$ is a monomial, and monomials can be ordered with a linear form $l : \mathbb{N}^{n+m} \to \mathbb{R}_+$. The whole theory then carries over to submodules of free modules, but here we only consider the case of ideals in S.

Definition. For $0 \neq f = \sum f_a x^a \in S$ let $\exp(f) = \max\{a \in \mathbb{N}^n \mid f_a \neq 0\}$, where the maximum is taken with respect to the order $<_l$ on $\mathbb{N}^n$. The *initial form* (or *leading term*) of f is $\mathrm{in}(f) := f_{\exp(f)} x^{\exp(f)}$. The *initial ideal* of an ideal $I \subset S$ is the ideal $\mathrm{in}(I) = \langle \mathrm{in}(f) \mid f \in I \setminus \{0\} \rangle$.

The geometry of the set $\exp(I) = \{\exp(f) \mid f \in I \setminus \{0\}\} \cup \{0\}$ reflects properties of the monomial ideal $\mathrm{in}(I)$. We note that $\exp(I) + \mathbb{N}^n = \exp(I)$, and that $\exp(I)$ has finitely many minimal elements with respect to the natural partial order on $\mathbb{N}^n$; they correspond to a finite set of generators of $\mathrm{in}(I)$.

Definition. A Gröbner basis of an ideal I is a set of generators $(f_1, \ldots, f_k)$ of I, such that their initial forms $(\mathrm{in}(f_1), \ldots, \mathrm{in}(f_k))$ generate $\mathrm{in}(I)$.

The condition in the definition that $(f_1, \ldots, f_k)$ generates I can be dropped, because it is a consequence of the fact that $(\mathrm{in}(f_1), \ldots, \mathrm{in}(f_k))$ generates $\mathrm{in}(I)$. For if $g \in I$, then $\mathrm{in}(g) \in \mathrm{in}(I)$, so $\mathrm{in}(g) = cx^a \mathrm{in}(f_j)$ for some j, some constant c, and a monomial x^a. Set $g' = g - cx^a f_j$, then $g' \in I$ and $g' = 0$ or $\exp(g') <_l \exp(g)$; in the latter case we repeat the procedure. After a finite number of steps the process stops. In fact, there exists a *division algorithm*: every $g \in S$ can be written as $g = \sum g_i f_i + h$, where all monomials in h belong to $\Delta(I)$, the set of all monomials not contained in the initial ideal $\mathrm{in}(I)$. The remainder $R(g) := h$ can be defined recursively by

$$R(g) = R(g - cx^a f_j) , \qquad \text{if } \mathrm{in}(g) = cx^a \mathrm{in}(f_j) \text{ for some } j ,$$

and

$$R(g) = \mathrm{in}(g) + R(g - \mathrm{in}(g)) , \qquad \text{if } \mathrm{in}(g) \notin \mathrm{in}(I) .$$

The algorithm can even be extended to a list $F = [f_1, \ldots, f_k]$ of polynomials, by setting $R_F(g) = R_F(g - cx^a f_j)$ for the smallest j such that $\mathrm{in}(g)$ is divisible by $\mathrm{in}(f_j)$, and in case no such j exists, by $R_F(g) = \mathrm{in}(g) + R_F(g - \mathrm{in}(g))$. Contrary to the case of a Gröbner basis, the outcome will in general depend on the order of the polynomials in the list.

The monomial order is often given in combinatorial terms. Two of the most used orders are the (graded) *lexicographic* order, defined by $x^a > x^b$ if and only if $\deg x^a > \deg x^b$ or the first non-zero entry in $a - b$ is positive, and the (graded) *reverse lexicographic* order: $x^a > x^b$ if and only if $\deg x^a > \deg x^b$ or the last non-zero entry in $a - b$ is negative; given a bound on the degree of polynomials, one can always find a linear form which induces these orders. In both the lexicographic and the reverse lexicographic order on $S = \mathbb{C}[w, x, y, z]$ we have that $w > x > y > z$, but the lexicographic order on the monomials of degree two is

$$w^2 > wx > wy > wz > x^2 > xy > xz > y^2 > yz > z^2 ,$$

while the reverse lexicographic order gives

$$w^2 > wx > x^2 > wy > xy > y^2 > wz > xz > yz > z^2 .$$

In [BM] BAYER and MUMFORD give an interpretation of the algorithm for Gröbner bases as computation of the specialisation from I to $\mathrm{in}(I)$. We explain this with the following example, taken from their paper.

Example. Let $S = \mathbb{C}[w, x, y, z]$ be the homogeneous coordinate ring of $\mathbb{P}^3$, and let $I \subset S$ be the ideal generated by

$$f_1 = w^2 - xy , \qquad f_2 = wy - xz , \qquad f_3 = wz - y^2 .$$

This ideal defines a rational normal curve X of degree 3. Consider the one parameter family λ_t, $t \neq 0$, of automorphisms of $\mathbb{P}^3$, defined by

$$\lambda_t \cdot (w, x, y, z) = (t^{-16} w, t^{-4} x, t^{-1} y, t^0 z) \,,$$

and let $X_t = \lambda_t(X) \cong X$. We want to determine the limit $X_0 = \lim_{t \to 0} X_t$. For this we take the scheme-theoretic closure $\overline{X}$ of the family X_t in $\mathbb{P}^3 \times \mathbb{C}$. Then the family is flat over the base space [Ha, III.9.8], so X_t is a deformation of X_0.

For $t \neq 0$ the curve X_t is defined by

$$g_1 = t^{32} \lambda_t \cdot f_1 = w^2 - t^{27} xy$$
$$g_2 = t^{17} \lambda_t \cdot f_2 = wy - t^{13} xz$$
$$g_3 = t^{16} \lambda_t \cdot f_3 = wz - t^{14} y^2 \,.$$

We remark that $\lambda \cdot w^a x^b y^c z^d = t^{-l} w^a x^b y^c z^d$ with $l = 16a + 4b + c$, so sorting monomials of degree < 4 by increasing powers of t for the action of λ_t is sorting them lexicographically. If we take the limit $t \to 0$ in (g_1, g_2, g_3), we get (w^2, wy, wz), which defines the plane $w = 0$ with an embedded point at $(0 : 1 : 0 : 0)$, so the fibre is too big. As flatness is equivalent to lifting of relations, we look at the syzygies on w^2, wy and wz. The module of syzygies is generated by the pairwise syzygies. Start with $(w^2)y - (wy)w = 0$. Substituting g_1 and g_2 for the lead terms, we get

$$(w^2 - t^{27} xy)y - (wy - t^{13} xz)w = t^{13} wxz - t^{27} xy^2 = g_3 t^{13} x \,.$$

Therefore the syzygy $g_1 y - g_2 w - g_3 t^{13} x = 0$ restricts for $t = 0$ to the monomial syzygy $(w^2)y - (wy)w = 0$. For $(w^2)z - (wz)z = 0$ we find the lift $g_1 z - g_2 t^{14} y - g_3 w = 0$. The third syzygy is more interesting:

$$(wy - t^{13} xz)z - (wz - t^{14} y^2)y = -t^{13} xz^2 + t^{14} y^3.$$

The lead term xz^2 does not lie in the ideal (w^2, wy, wz). The solution to our problem is to add a generator $g_4 = xz^2 - ty^3$ to the ideal; for $t = 1$ this means taking the extra generator $f_4 = xz^2 - y^3$. For $t \neq 0$ this does not change the ideal, but it cuts away the extra component over $t = 0$. We then have the syzygy $g_2 z - g_3 y + g_4 t^{13} = 0$. The syzygy $(wz)xz - (xz^2)w$ lifts to

$$g_2(-ty^2) + g_3(xz) - g_4 w = 0.$$

This shows that the ideal (g_1, g_2, g_3, g_4) defines a flat family; the set of generators

$$f_1 = w^2 - xy \,, \qquad f_2 = wy - xz \,, \qquad f_3 = wz - y^2 \,, \qquad f_4 = xz^2 - y^3$$

is a Gröbner basis for I, and the limit X_0 is given by the ideal $\text{in}(I) = (w^2, wy, wz, xz^2)$. It consists of the reduced line $w = x = 0$, the line $w = z = 0$ with a double structure, and an embedded point at $(0 : 1 : 0 : 0)$. The equation $xz^2 - y^3 = 0$ defines the projection of X into the plane $w = 0$.

Consider now the same ideal, but with action

$$\lambda_t \cdot (w, x, y, z) = (t^0 w, t^1 x, t^4 y, t^{16} z) \, ,$$

giving the reverse lexicographic order. For $t \neq 0$ the curve X_t is defined by

$$
\begin{aligned}
g_1 &= \lambda_t \cdot f_1 &&= w^2 - t^5 xy \\
g_2 &= t^{-4} \lambda_t \cdot f_2 &&= wy - t^{13} xz \\
g_3 &= t^{-8} \lambda_t \cdot (-f_3) &&= y^2 - t^8 wz \, .
\end{aligned}
$$

In the limit $t \to 0$ we get the ideal (w^2, wy, y^2), which defines the line $w = y = 0$ with a multiplicity three structure. This family is already flat: the module of syzygies is generated by $g_1 y - g_2 w + g_3 t^5 x = 0$ and $g_1 t^8 z + g_2 y - g_3 w = 0$. The set of generators $(w^2 - xy, wy - xz, y^2 - wz)$ is a Gröbner basis.

This example illustrates that the dominant effect of the lexicographic order is a projection from $\mathbb{P}^{n-1}$ to $\mathbb{P}^{n-2}$ from the point $(1 : 0 : \dots : 0)$, i.e., to eliminate x_1. The second order effect is to project onto $\mathbb{P}^{n-3}$, and so on. For the reverse lexicographic order the dominant effect is a projection onto the last coordinate point $(0 : \dots : 0 : 1)$, with as second order effect a projection on the last coordinate line; this order tries first to make X into a cone over $x_n = 0$. Projections change ideals much more drastically than linear sections; therefore the reverse lexicographic order is preferable in many circumstances — in the computer algebra system *Macaulay* [BS] it is the default.

In practice one forgets about the parameter t and works directly in the ring S. Let f and g be two polynomials, with leading terms $\text{in}(f) = f_a x^a$ and $\text{in}(g) = g_b x^b$. Define the *S-polynomial* of f and g as

$$S(f, g) = g_b x^{M-a} f - f_a x^{M-b} g \, ,$$

where $M = \sup(a, b)$, or in other words x^M is the least common multiple of x^a and x^b. By construction $\exp(S(f, g)) < M$. If f and g are polynomials in a list $F = [f_1, \dots, f_k]$, then computing the rest $R_F(S(f, g))$ just means trying to lift the syzygy $\text{in}(f) \cdot g_b x^{M-a} - \text{in}(g) \cdot f_a x^{M-b}$, and if the rest is zero, a syzygy between the f_i is found. If not, we add $R_F(S(f, g))$ to our list, and form new S-polynomials. Because the ring S is Noetherian, after a finite number of additions all rests are zero.

Proposition (Buchberger). *If $F = [f_1, \dots, f_k]$ is a list of polynomials with the property that $R_F(S(f_i, f_j)) = 0$ for all $i < j$, then F is a Gröbner basis for the ideal $I = (f_1, \dots, f_k)$.*

Proof. If $x^a \in \text{in}(I)$, then $x^a = \text{in}(g)$ for some $g = \sum \lambda_i f_i$. Let $m = \max_i \{ \exp(\lambda_i f_i) \}$. If $m = a$, we are done. Otherwise we decrease m by adding syzygies $\sum f_i r_i = 0$ to $g = \sum \lambda_i f_i$; this does not change g. More precisely, let i be the smallest index, such that $\exp(\lambda_i f_i) = m$; as x^m is not a monomial of g, the term $\text{in}(\lambda_i f_i)$ cancels, and there exists a j with $i < j$ and $\exp(\lambda_j f_j) = m$. The S-polynomial $S(f_i, f_j)$ leads to a syzygy $S_{ij} = \sum f_\alpha r_\alpha$, with $\exp(f_i r_i) = \exp(f_j r_j) \leq m$, so $\text{in}(\lambda_i f_i) = c x^k \text{in}(f_i r_i)$ for some $k \in \mathbb{N}$, and $\exp(f_\alpha r_\alpha) < \exp(f_i r_i)$ for $\alpha \neq i, j$. Now write $g = \sum_k \lambda_k f_k - c x^k S_{ij}$. Proceeding in this way we eventually decrease m. $\qquad \square$

To give the definitions in the power series case we let S now be the ring $\mathbb{C}\{x_1, \ldots, x_n\}$. The only thing to be changed in the definition of initial form is to replace the maximum by the minimum. Alternatively, we replace l by $-l$, so we work with an injective linear form $l \colon \mathbb{N}^n \to \mathbb{R}_-$.

Definition. A standard basis of an ideal I is a set of generators $(f_1, \ldots, f_k)$ of I, such that their initial forms $(\mathrm{in}(f_1), \ldots, \mathrm{in}(f_k))$ generate $\mathrm{in}(I)$.

The existence follows from the generalised Weierstraß division theorem, see [Gal]. Let I be an ideal in $S = \mathcal{O}_n$, and define

$$\Delta(I) = \{ f = \sum f_a x^a \in S \mid f_a = 0 \text{ if } a \in \exp(I) \} .$$

We quote the following theorem.

Theorem. *The natural projection $\Delta(I) \to S/I$ is an isomorphism of vector-spaces. Equivalently, given a set of generators $(f_1, \ldots, f_k)$ of I, then for every $g \in S$ there exist $q_i \in S$ and a unique $r \in \Delta(I)$ such that $g = qf + r$.*

Corollary. *One may choose generators f_i of I such that $f_i - \mathrm{in}(f_i) \in \Delta(I)$.*

Proof (of the Corollary). Let g_j be any set of generators, whose initial terms generate $\mathrm{in}(I)$. Divide $g_j - \mathrm{in}(g_j)$ by the g_i's:

$$g_j - \mathrm{in}(g_j) = \sum q_{ji} g_i + r_j ,$$

and put $f_j = \mathrm{in}(g_j) + r_j$. Let $J \subset I$ be the ideal, generated by the f_j. Then $\Delta(J) = \Delta(I)$, so by the division Theorem $I = J$. □

The division algorithm from the polynomial case does not carry over, as can be seen from a simple example in one variable: let $I \subset \mathbb{C}\{x\}$ be the ideal generated by $x - x^2$ and consider the polynomial x. Then $\mathrm{in}(x) = \mathrm{in}(x - x^2)$, so a first reduction gives $x \equiv x^2 \pmod{I}$, and continuing we get the never ending process $x \equiv x^2 \equiv x^3 \equiv x^4 \equiv \cdots$.

Standard bases can be computed with MORA's tangent cone algorithm. A generalisation is implemented in the system *Singular* [GPS]. Of course one can only compute with polynomials. The ring involved is a suitable localisation of the polynomial ring. In the simple example one has $x = (1-x)^{-1}(x - x^2)$, which one has to use in the form $x \cdot (1 - x) = x - x^2$.

One defines a *weighted tangent cone* with a linear form $l \colon \mathbb{N}^n \to \mathbb{R}_+$, in general not injective. For $f = \sum f_a x^a \in S$ we define the l-degree of f as $m(f) = \min\{l(a) \mid f_a \neq 0\}$; the initial form of f is $\mathrm{in}(f) := \sum_{\{a : l(a) = m\}} f_a x^a$. As before, a standard basis of an ideal I is a set of generators, whose initial forms generate the ideal $\mathrm{in}(I)$. For $l(a) = a_1 + \cdots + a_n$ the ideal $\mathrm{in}(I)$ gives the usual tangent cone. In general the ideal $\mathrm{in}(I)$ is not a monomial ideal. The existence of a standard basis follows from the monomial case: for a total order, refining the partial order defined by l, the (monomial) ideal of initial forms of I and of $\mathrm{in}(I)$ coincide, so a standard basis of I also generates $\mathrm{in}(I)$.

Proposition. *In this situation the ideal I is a deformation of $\mathrm{in}(I)$.*

Proof. Consider the diagonal action of $\mathbb{C}^*$ on S, given by $x_i \mapsto t^{l_i} x_i$. Let $(f_1, \ldots, f_k)$ be a standard basis of I, and let m_j be the l-degree of f_j. Consider the family, given by $g_j(x) = t^{-m_j} f(t^l x) = \mathrm{in}(f_j(x)) + t f'_j(x, t)$. The family is flat, because $(f_1, \ldots, f_k)$ is a standard basis. $\qquad\square$

Corollary. *Every singularity is a deformation of its tangent cone.*

We use the Weierstraß Division Theorem for another interpretation of flatness. Realise as before the map $\pi: (X, 0) \to (S, 0)$ via an embedding $(X, 0) \hookrightarrow (S \times \mathbb{C}^N, 0)$. Then X is given by an ideal I in $\mathcal{O}_S\{x_1, \ldots, x_N\} = \mathcal{O}_{S,0} \widehat{\otimes} \mathbb{C}\{x_1, \ldots, x_N\}$, and X_0 is defined by the ideal I_0. Given a linear form $l: N^n \to \mathbb{R}_+$, Weierstraß Division yields an isomorphism as $\mathbb{C}$-vector spaces between $\mathcal{O}_{X_0}$ and

$$\Delta(I_0) = \{f = \sum f_a x^a \in \mathbb{C}\{x\} \mid f_a = 0 \text{ if } a \in \exp(I_0)\} \ .$$

This means that we have a precise description of what $\mathcal{O}_{X_0}$ looks like, if we are willing to forget the ring structure. By a generic choice of coordinates we can bring $\Delta(I_0)$ in a particular nice form, following GRAUERT [Gra], cf. [Gal, Hi2]. As ordering we choose the lexicographic order: $x^a < x^b$ if and only if $|a| < |b|$ or the first non-zero entry in $a - b$ is positive. This has the effect that $x_1 < \cdots < x_N$. To understand the behaviour of the set of exponents for this order under coordinate transformations, it suffices to look at linear transformations; we may also suppose that I_0 is homogeneous: if we temporarily denote by $\mathrm{in}_h(f)$ the homogeneous term of lowest degree (the initial form for $l(a) = a_1 + \cdots + a_N$), and write as usual $\exp(f)$ for the exponent in the lexicographic order, then $\mathrm{in}_h(\sigma^* I_0) = (d\sigma)^*(\mathrm{in}_h(I_0))$, and as $\exp(I_0) = \exp(\mathrm{in}_h(I_0))$, we find that $\exp(\sigma^* I_0) = \exp((d\sigma)^*(\mathrm{in}_h(I_0)))$, where σ is any coordinate transformation.

Proposition. *For a Zariski open subset $U \subset Gl_N$ the set $E = \exp(\sigma^*(I_0))$ is independent of $\sigma \in U$. It is monotonous: for all $a = (a_1, \ldots, a_N) \in E$ and $i < N$ also*

$$(a_1, \ldots, a_{i-1}, a_i + a_{i+1}, 0, a_{i+2}, \ldots, a_N) \in E \ .$$

There exist finite sets F_i, $1 \le i \le N$, such that

$$\mathbb{N}^N \setminus E = \cup_{i=1}^N F_i \times \mathbb{N}^{N-i} \ .$$

Proof. We construct E and U by induction: suppose generators $a^1 < \ldots < a^i$ of E are found, and a Zariski open set U_i, such that for all $\sigma \in U_i$ the a^j are the first i generators of $\exp(\sigma^*(I_0))$; fix a $\sigma_0 \in U_i$. Take a^{i+1} as the minimum $(i+1)$-st generator over all $\sigma \in U_i$: take a standard basis for $\sigma_0^*(I_0)$, apply a transformation $\sigma = \sigma' \circ \sigma_0 \in U_i$, and construct a standard basis for $\sigma^*(I_0)$; the coefficients of the first $i + 1$ elements $f_1, \ldots, f_{i+1}$, are rational functions of the entries of σ', so the minimal value is obtained for a Zariski open subset $U_{i+1} \subset U_i$.

Now we show that E is monotonous. Let $a \in E$, so there exists an $f \in I_0$ with $\exp(f) = a$. Consider the coordinate transformation σ, given by $x'_{i+1} = x_{i+1} + \lambda x_i$, and $x_j = x'_j$ for $j \neq i+1$. Then $\exp(f(x')) = (a_1, \ldots, a_{i-1}, a_i + a_{i+1}, 0, a_{i+2}, \ldots, a_N)$.

Let $\mathrm{pr}_i \colon \mathbb{N}^N \to \mathbb{N}^i$ be the projection on the first i coordinates. Define $F_i = (\mathrm{pr}_{i-1}(E) \times \mathbb{N}) \setminus \mathrm{pr}_i(E)$. Then $\mathbb{N}^N \setminus E = \cup_{i=1}^N F_i \times \mathbb{N}^{N-i}$. Suppose F_i is not finite, then a sequence $(a(k))_{k \in \mathbb{N}}$ exists with $a(k) \in \mathrm{pr}_{i-1} E$, but $(a(k), 0) \notin \mathrm{pr}_i E$, and at least one coefficient $a_j(k)$ unbounded. By taking subsequences we may assume that all $a_j(k)$ are either constant or strictly increasing. As $a \in \mathrm{pr}_i E$ implies that $a + \mathbb{N}^i \in \mathrm{pr}_i E$, we conclude the existence of a sequence $a(k) \in \mathrm{pr}_{i-1} E$, but $(a(k), 0) \notin \mathrm{pr}_i E$ with exactly one coefficient $a_j(k)$ unbounded. Because $a(1) \in \mathrm{pr}_{i-1} E$, there exists a number α such that $(a(1), \alpha) \in \mathrm{pr}_i E$. As $\mathrm{pr}_i E$ is monotonous, this implies that $(a_1(1), \ldots, a_{j-1}(1), A_j, 0, \ldots, 0) \in \mathrm{pr}_i E$ for some large number A_j, contradicting the fact that $(a(k), 0) \notin \mathrm{pr}_i E$ for all k. □

Now define for $(X, 0) \hookrightarrow (S \times \mathbb{C}^N, 0)$:

$$\Delta(I) = \left\{ f = \sum f_a(t) x^a \in \mathcal{O}_S\{x\} \mid f_a = 0 \text{ if } a \in \exp(I_0) \right\} .$$

Theorem. *The natural projection $\Delta(I) \to \mathcal{O}_S\{x\}/I$ is surjective. It is bijective if and only if the map $\pi \colon X \to S$ is flat.*

Sketch of proof. The first statement is the Division Theorem with parameters (cf. [Gal]); in this case one can also derive it from the Division Theorem by lifting a standard basis of I_0 to a set of elements of I, and applying the Theorem to the ideal in $\mathbb{C}\{x, t\}$, generated by these functions, with a suitable chosen linear form, cf. [Hi2]. The converse follows from the lemma on flatness in Chap. 1. □

In the flat case the algebra $\Delta(I)$ is a free $\mathcal{O}_S$-module, so in a deformation the module structure does not change, *we only alter the algebra structure.*

Example: triple covers. Let $\pi \colon (Y, 0) \to (X, 0)$ be a finite map of degree three. Then π can be regarded as a flat deformation of the fibre Y_0, which is a fat point of multiplicity three. There are two types of fat points, the complete intersection $y^3 = 0$, and $\mathcal{O}_2/\mathfrak{m}^2$, with equations $y^2 = yz = z^2 = 0$. In the first case the local ring has as $\mathbb{C}$-basis $\{1, y, y^2\}$ — if we put $z = y^2$, we obtain the same basis $\{1, y, z\}$ as in the second case. Therefore we have an additive realisation of $\mathcal{O}_Y$ as $\mathcal{O}_X$-module:

$$\mathcal{O}_Y \cong \mathcal{O}_X + \mathcal{O}_X \cdot y + \mathcal{O}_X \cdot z .$$

To describe the algebra structure, we have to give the values of the following products:

$$y^2 = a_0(x) + a_1(x)y + a_2(x)z$$
$$yz = b_0(x) + b_1(x)y + b_2(x)z$$
$$z^2 = c_0(x) + c_1(x)y + c_2(x)z .$$

Commutativity is implicit in these definitions: we put $zy = yz$. The fact that the multiplication has to be associative gives non-trivial conditions on the coefficients

$$0 = z(y^2) - y(yz) = z(a_0 + a_1 y + a_2 z) - y(b_0 + b_1 y + b_2 z) =$$
$$a_1(b_0 + b_1 y + b_2 z) + a_2(c_0 + c_1 y + c_2 z) - b_1(a_0 + a_1 y + a_2 z)$$
$$- b_2(b_0 + b_1 y + b_2 z) + a_0 z - b_0 y = (a_2 c_1 - b_2 b_1 - b_0)y$$
$$+ (a_1 b_2 + a_2 c_2 - b_1 a_2 - b_2^2 + a_0)z + a_1 b_0 + a_2 c_0 - b_1 a_0 - b_2 b_0 ,$$
$$0 = z(yz) - y(z^2) = z(b_0 + b_1 y + b_2 z) - y(c_0 + c_1 y + c_2 z) =$$
$$b_1(b_0 + b_1 y + b_2 z) + b_2(c_0 + c_1 y + c_2 z) - c_1(a_0 + a_1 y + a_2 z)$$
$$- c_2(b_0 + b_1 y + b_2 z) + b_0 z - c_0 y = (b_1^2 + b_2 c_1 - c_1 a_1 - c_2 b_1 - c_0)y$$
$$+ (b_1 b_2 - c_1 a_2 + b_0)z + b_1 b_0 + b_2 c_0 - c_1 a_0 - c_2 b_0 .$$

This gives
$$a_0 = b_2^2 + b_1 a_2 - a_1 b_2 - a_2 c_2$$
$$b_0 = a_2 c_1 - b_2 b_1$$
$$c_0 = b_1^2 + b_2 c_1 - c_1 a_1 - c_2 b_1 .$$

In deriving these equations we used the formulas for the products to rewrite expressions; another way to look at it is that we found syzygies between these equations. They give the following determinantal description of Y:

$$\text{Rank} \begin{pmatrix} z - b_1 & -y + a_1 - b_2 & a_2 \\ -c_1 & z + b_1 - c_2 & -y + b_2 \end{pmatrix} \leq 1 .$$

3 Infinitesimal deformations

Let $(X,0) \subset (\mathbb{C}^N, 0)$ be an analytic germ, with local ring $\mathcal{O}_{X,0}$. In this section we always consider germs at the origin, and therefore we drop the 0 from the notation. The ring $\mathbb{C}\{x\}$ of power series in N variables will be denoted by $\mathcal{O}_N$. The first few terms of the resolution of $\mathcal{O}_X$ are

$$0 \longleftarrow \mathcal{O}_X \longleftarrow \mathcal{O}_N \xleftarrow{f} \mathcal{O}_N{}^k \xleftarrow{r} \mathcal{O}_N{}^l .$$

The entries of the row vector $f = (f_1, \dots, f_k)$ generate the ideal I of X, and the columns of r generate the relations.

A *first order infinitesimal deformation* of X is a deformation over the double point $\mathbb{D}$, the zero-dimensional space with as local ring the ring of dual numbers $\mathbb{C}[t]/(t^2)$. We also write this ring as $\mathbb{C}[\varepsilon]$, where ε is defined as variable with the property that $\varepsilon^2 = 0$. So let $\mathcal{X} \to \mathbb{D}$ be a deformation of X. Then there is a resolution

$$0 \longleftarrow \mathcal{O}_{\mathcal{X}} \longleftarrow \mathcal{O}_{\mathbb{C}^N \times \mathbb{D}} \xleftarrow{F} \mathcal{O}_{\mathbb{C}^N \times \mathbb{D}}^k \xleftarrow{R} \mathcal{O}_{\mathbb{C}^N \times \mathbb{D}}^l ,$$

with $F = f + \varepsilon f'$ and $R = r + \varepsilon r'$. As $\varepsilon^2 = 0$, the condition $FR = 0$ gives

$$FR = (f + \varepsilon f')(r + \varepsilon r') = fr + \varepsilon(fr' + f'r) = 0 .$$

Because $fr = 0$, we obtain the equation $fr' + f'r = 0$ in $\mathcal{O}_N$.

The first order deformations form an $\mathcal{O}_X$-module: if $(f + \varepsilon f'_1)(r + \varepsilon r'_1) = 0$ and $(f + \varepsilon f'_2)(r + \varepsilon r'_2) = 0$, then $\big(f + \varepsilon(f'_1 + f'_2)\big)\big(r + \varepsilon(r'_1 + r'_2)\big) = fr + \varepsilon\big(f(r'_1 + r'_2) + (f'_1 + f'_2)r\big) = 0$. Furthermore, $(f + \varepsilon\phi f')(r + \varepsilon\phi r') = 0$ for $\phi \in \mathcal{O}_N$. Finally, if $f' \in I^k \subset \mathcal{O}_N{}^k$, then there exists a matrix $M \in M_k(\mathcal{O}_N)$ with $f + \varepsilon f' = f(\mathrm{Id} + \varepsilon M)$; as $\mathrm{Id} + \varepsilon M$ is invertible (with inverse $\mathrm{Id} - \varepsilon M$), the ideals generated by f and by $f + \varepsilon f'$ are equal.

Proposition. *The $\mathcal{O}_X$-module of first order deformations is isomorphic to the normal module $N_X = \mathrm{Hom}_{\mathcal{O}_X}(I/I^2, \mathcal{O}_X)$.*

Proof. Let $f + \varepsilon f'$ be an infinitesimal deformation. Then f' determines an $\mathcal{O}_N$-homomorphism $\mathcal{O}_N{}^k \to \mathcal{O}_N$, which maps the image of r into I, because $f'r = -fr'$. Hence f' induces a homomorphism $\rho(f')\colon \mathcal{O}_N{}^k/\mathrm{Im}\,r = I \longrightarrow \mathcal{O}_N/I = \mathcal{O}_X$, which sends f_i to $(f'_i \bmod I)$. This means that $\rho(f')$ is an element of $\mathrm{Hom}_{\mathcal{O}_N}(I, \mathcal{O}_X) \cong \mathrm{Hom}_{\mathcal{O}_X}(I/I^2, \mathcal{O}_X)$.

Conversely, given a homomorphism $\varphi \in \operatorname{Hom}_{\mathcal{O}_N}(I, \mathcal{O}_X)$, we lift the vector

$$\varphi(f) = (\varphi(f_1), \ldots, \varphi(f_k)) \in \mathcal{O}_X{}^k$$

to a vector $f' \in \mathcal{O}_N{}^k$, inducing a homomorphism $\tilde{\varphi} \colon \mathcal{O}_N{}^k \to \mathcal{O}_N$. For every relation r_j the function $f' r_j = \tilde{\varphi}(r_j)$ is a lift of $\varphi(\sum f_i r_{ij}) = 0 \in \mathcal{O}_X$. Therefore one can find a matrix r' with $f' r + f r' = 0$. Any two liftings of $\varphi(f)$ differ by a $g \in I^k$, so they determine the same deformation. □

An infinitesimal deformation $f + \varepsilon f'$ is trivial, if there is an automorphism $\varphi(x, \varepsilon) = (x + \varepsilon \delta(x), \varepsilon)$ of $\mathbb{C}^N \times \mathbb{D}$, such that $f + \varepsilon f'$ and $f \circ \varphi$ determine the same ideal. Let Θ_N be module of germs of vector fields at the origin. The computation

$$\frac{d}{d\varepsilon} f \circ \varphi(x, \varepsilon)|_{\varepsilon = 0} = \frac{d}{d\varepsilon} f(x + \varepsilon \delta(x))|_{\varepsilon = 0} = \sum_j \frac{\partial f}{\partial x_j} \delta_j(x) \ .$$

shows that the trivial deformations are the image of the natural map

$$\Theta_N|_X = \Theta_N \otimes \mathcal{O}_X \longrightarrow \operatorname{Hom}_{\mathcal{O}_N}(I, \mathcal{O}_X) = N_X \ ,$$

which sends a vector field δ to the homomorphism $g \mapsto \delta(g)$. The kernel of this map is the $\mathcal{O}_X$-module $\Theta_X = \{\delta|_X \mid \delta(I) \subset I\}$. One has $\Theta_X = \operatorname{Hom}_X(\Omega_X^1, \mathcal{O}_X)$.

Definition. The module T_X^1 of isomorphism classes of first order infinitesimal deformations is

$$T_X^1 = \operatorname{coker}\{\Theta_N|_X \to N_X\} \ .$$

Example: hypersurface singularities. In this case the ideal I is principal, generated by a function f, and $N_X = \operatorname{Hom}(I/I^2, \mathcal{O}_X)$ is a free $\mathcal{O}_X$-module with $f \mapsto 1$ as generator. Therefore

$$T_X^1 = \mathcal{O}_{n+1} / \left(f, \frac{\partial f}{\partial x_0}, \ldots, \frac{\partial f}{\partial x_n} \right) \ .$$

Remark. If the germ X is smooth, then $T_X^1 = 0$. In fact, we may assume that we have equality of germs $(\mathbb{C}^N, 0) = (X, 0)$, so $N_X = 0$. In the algebraic situation, for affine X, this argument does not work, but it is well known that for a nonsingular subvariety X of a nonsingular variety Y the sequence of sheaves

$$0 \longrightarrow \Theta_X \longrightarrow \Theta_Y \otimes \mathcal{O}_X \longrightarrow N_X \longrightarrow 0$$

is exact [Ha, p. 182].

Until now we worked with modules over the local ring at the origin. Our definitions extend to give a coherent sheaf T_X^1 on an analytic space X. If the germ $(X, 0)$ has an isolated singularity, then by coherence of this sheaf T_X^1 is a finite-dimensional vector space.

Remark. With a computer algebra package, which can compute standard bases and syzygies, one can compute T^1 for a given singularity. The first step is to determine the $\mathcal{O}_X$-module N_X. Every generator $f + \varepsilon f'$ satisfies $f'r \equiv 0$ (mod I). Taking the transpose gives ${}^t r {}^t f' \equiv 0$ (mod I); in other words, ${}^t f'$ is a syzygy between the columns of the matrix ${}^t r$ over the ring $\mathcal{O}_X$. Therefore, to find N_X one has to find the syzygy module of ${}^t r$ in $\mathcal{O}_X$. The partial derivatives of the row vector f generate a submodule of N_X, and one has to determine the quotient.

The vector space T_X^2 comes into play, if we try to extend first order solutions of the equations $FR = 0$ to higher order. For notational convenience we consider one parameter deformations: let $\mathcal{X}$ be a deformation of X over the fat point with local ring $A = \mathbb{C}[t]/(t^n)$, so we have F_{n-1} and R_{n-1} with $F_{n-1} R_{n-1} \equiv 0$ (mod t^n). We want a solution modulo t^{n+1}: a deformation over $A' = \mathbb{C}[t]/(t^{n+1})$. Suppose we have a solution $F_n = F_{n-1} + t^n f^{(n)}$, $R_n = R_{n-1} + t^n r^{(n)}$, then

$$F_n R_n \equiv F_{n-1} R_{n-1} + t^n \left(f^{(n)} R_{n-1} + F_{n-1} r^{(n)} \right)$$
$$\equiv F_{n-1} R_{n-1} + t^n \left(f^{(n)} r + f r^{(n)} \right) \equiv 0 \pmod{t^{n+1}} .$$

Because $F_{n-1} R_{n-1} \equiv 0$ (mod t^n), we have in $\mathcal{O}_X{}^l$ the equation

$$t^{-n} F_{n-1} R_{n-1} + f^{(n)} r \equiv 0 \pmod{t} .$$

To find a lift F_n we have to solve this equation for $f^{(n)}$.

Lemma. *The vector $t^{-n} F_{n-1} R_{n-1} \bmod t \in \mathcal{O}_X{}^l$ depends only on r (and F_{n-1}), but not on the lift R_{n-1} of r. It represents an element of the vector space $\mathrm{Hom}_{\mathcal{O}_N}(\mathcal{O}_U{}^l, \mathcal{O}_X)$, which vanishes on $\ker r$. Let $\mathcal{R} = \mathcal{O}_N{}^l / \ker r$ be the module of relations, and let $\mathcal{R}_0$ be the submodule of Koszul relations. Then $\mathcal{R}/\mathcal{R}_0$ is an $\mathcal{O}_X$-module. The vector $t^{-n} F_{n-1} R_{n-1}$ descends to an element of $\mathrm{Hom}(\mathcal{R}/\mathcal{R}_0, \mathcal{O}_X)$.*

Proof. In the statement of the Lemma F_{n-1} denotes any lift to $A'\{x\}$ of $F_{n-1} \in A\{x\}$. An obvious choice is a lift which does not involve terms with t^n. The difference between two lifts is of the form $g^{(n)} r$.

Suppose $F_{n-1}(r + tR) \equiv F_{n-1}(r + tR') \equiv 0$ (mod t^n), then $t^{-1} F_{n-1}(R - R') \equiv 0$ (mod t^{n-1}). Thus $t^{-1}(R - R') \bmod t^{n-1}$ is a relation, which can be extended to order $n - 1$: there exists R'' with

$$F_{n-1}\left(t^{-1}(R - R') + t^{n-1} R'' \right) \equiv 0 \pmod{t^n} ,$$

and therefore

$$F_{n-1}\left((R - R') + t^n R'' \right) \equiv 0 \pmod{t^{n+1}} .$$

As $t^n F_{n-1} R'' \equiv t^n f R''$ (mod t^{n+1}), we obtain that the classes of $F_{n-1} \cdot (r + tR)$ and $F_{n-1} \cdot (r + tR')$ in $\mathcal{O}_X{}^l$ are the same. In a similar way one shows that all the components of the vector $t^{-n} F_{n-1} R_{n-1} v$ lie in I if $v \in \ker r$.

The value of the homomorphism on any relation can be computed by taking a suitable lift of the particular relation. The canonical choice for a Koszul relation r' gives $F_{n-1}R' = 0 \in \mathcal{O}_N$.   □

The equation $t^{-n}F_{n-1}R_{n-1} + f^{(n)}r \equiv 0$ can be solved if and only if the homomorphism $t^{-n}F_{n-1}R_{n-1}$ can be lifted to the element of $\operatorname{Hom}(\mathcal{O}_N{}^k, \mathcal{O}_X)$, given by the vector $f^{(n)}$. Therefore one defines

$$T_X^2 = \operatorname{coker}\{\operatorname{Hom}(\mathcal{O}_N{}^k, \mathcal{O}_X) \longrightarrow \operatorname{Hom}(\mathcal{R}/\mathcal{R}_0, \mathcal{O}_X)\}\,.$$

Example. For complete intersections $\mathcal{R} = \mathcal{R}_0$, so $T^2 = 0$.

The previous considerations are easily adapted to handle more general infinitesimal base spaces. The result is the following:

Proposition. *Let* $0 \to J \to A' \to A \to 0$ *be a small extension of Artinian rings (which means that* $J^2 = 0$*). The obstruction to extend a deformation* $\mathcal{X}_A$ *of* X *over* A *to a deformation over* A' *lies in* $T_X^2 \otimes J$.

Remark. Suppose the singularity X admits a good $\mathbb{C}^*$-action, so $\mathcal{O}_X$ is a (positively) graded module. Then all modules T_X^i inherit a grading.

Theorem [Pin1]. *A singularity* X *with good* $\mathbb{C}^*$ *-action has a* $\mathbb{C}^*$ *-equivariant miniversal deformation* $\pi \colon \mathcal{X} \to S$. *The restriction* $\pi_- \colon \mathcal{X}_- \to S_-$ *to the subspace of negative weight is versal for deformations of* X *with negative weight.*

For the precise definition of versality I refer to Chapter 6. The restricted deformation $\pi_- \colon \mathcal{X}_- \to S_-$ can be defined by weighted homogeneous polynomials, which can be computed in a finite number of steps.

We illustrate the previous general theory with a computation for a simple class of curve singularities.

Example: L_n^n. Consider the singularity $X = L_n^n$, consisting of the n coordinate axes in $\mathbb{C}^n$, with $n \geq 3$. It is given by $\binom{n}{2}$ equations $f_{ij} = x_i x_j$, $1 \leq i < j \leq n$, and the relations among them are $f_{ij}x_k - f_{ik}x_j = 0$; denote this relation by r_{ijk}. For each triple (i, j, k) of distinct indices one can form three such relations, but only two are linearly independent. Therefore the minimal number of generators of the module of relations is $2\binom{n}{3}$, the same number we found for the cone over the rational normal curve of degree n. Indeed, the general hyperplane section of such a cone is the curve formed by n lines in general position, which is isomorphic to the coordinate axes.

We first determine the normal module N_X. Deform the equations to $F_{ij} = f_{ij} + f'_{ij}$. The condition that this defines a first order deformation of X is that $f'_{ij}x_k - f'_{ik}x_j = 0$ in $\mathcal{O}_X$; reducing this equation modulo $\mathfrak{m}_X^2$, we see that the constant terms $f'_{ij;0}$ satisfy $f'_{ij;0}x_k - f'_{ik;0}x_j = 0$ and therefore $f'_{ij;0} = 0$ for all $i < j$. As the maximal ideal of X is equal to $x_1\mathbb{C}\{x_1\} \oplus \cdots \oplus x_n\mathbb{C}\{x_n\}$, we can write

$$F_{ij} = x_i x_j - \sum_k x_k a_{ij}^k(x_k)\,,$$

with $a_{ij}^k(x_k)$ a power series in x_k. In $\mathcal{O}_X$ we compute $F_{ij}x_k - F_{ik}x_j = x_j^2 a_{ik}^j(x_j) - x_k^2 a_{ij}^k(x_k)$, so $a_{ij}^k = 0$ for $k \neq i,j$. Therefore N_X is generated by $n(n-1)$ elements e_{ij}; the corresponding first order deformation is given by

$$F_{ij} = x_i x_j - \varepsilon x_j, \qquad F_{kl} = x_k x_l, \quad \text{if } \{k,l\} \neq \{i,j\} \ .$$

Note that $x_k e_{ij} = 0$ in N_X for $k \neq j$.

The deformation e_{ij} extends to a deformation over a one-dimensional base, essentially given by the same formula: just replace ε by t; it moves the x_j-axis along the x_i-axis, and leaves the other lines fixed.

We compute the image of $\rho \colon \Theta_U|_X \to N_X$. As $\frac{\partial}{\partial x_k} f_{ij} = \delta_{ki}x_j + \delta_{kj}x_i$, we have that $\rho(\frac{\partial}{\partial x_k}) = -\sum_{i \neq k} e_{ki}$. Furthermore $\rho(x_i \frac{\partial}{\partial x_k}) = -x_i e_{ki}$, so $\rho(\mathfrak{m}_X \Theta_U|_X) = \mathfrak{m}_X N_X$. We have shown:

Lemma. *The dimension of T_X^1 is $n(n-2)$.*

In describing T_X^1 it is important to preserve the symmetry of the equations. Therefore we realise T^1 as linear subspace of $\mathbb{C}^{n(n-1)}$: take $n(n-1)$ coordinates a_{ij}, where $i \neq j$, subject to n linear relations which we can take as $\sum_j a_{ij} = 0$. We now write

$$F_{ij} = x_i x_j - a_{ij}x_j - a_{ji}x_i \ .$$

Let $A = \mathbb{C}\{a_{12}, \ldots, a_{n-1,n}\}$. The equations F_{ij} define a deformation of X over $A/\mathfrak{m}_A^2$ (so for the moment we are forgetting about the condition $\sum_j a_{ij} = 0$). We may as well write

$$F_{ij} = (x_i - a_{ij})(x_j - a_{ji}) \ .$$

We lift the relations r_{ijk} and compute

$$(x_k - a_{ki})F_{ij} - (x_j - a_{ji})F_{ik} + (a_{ij} - a_{ik})F_{jk}$$
$$\equiv x_k(a_{ik} - a_{ij})(a_{jk} - a_{ji}) - x_j(a_{ij} - a_{ik})(a_{kj} - a_{ki}) \pmod{\mathfrak{m}_A^3} \ .$$

The right-hand side can be made to vanish by changing F_{ij} into

$$F_{ij} = (x_i - a_{ij})(x_j - a_{ji}) - (a_{ik} - a_{ij})(a_{jk} - a_{ji}) \ ,$$

and F_{ik} in a similar way, with j and k interchanged. In order to lift simultaneously the other relations, like r_{ijl} with $l \neq k$, we require that the new summand in F_{ij} is independent of k:

$$(a_{ik} - a_{ij})(a_{jk} - a_{ji}) - (a_{il} - a_{ij})(a_{jl} - a_{ji}) = 0 \ .$$

This gives $n(n-1)(n-3)/2$ linear independent equations, but to choose them from the $\binom{n}{2}\binom{n-2}{2}$ above means breaking the symmetry of the equations. These equations define a space $S \subset \mathbb{C}^{n(n-1)}$, over which the F_{ij} describe a flat deformation of X: one checks easily that $(x_k - a_{ki})F_{ij} - (x_j - a_{ji})F_{ik} +$

$(a_{ij} - a_{ik})F_{jk} = 0$, if only the indices i, j and k are used, and the equations for the base space allow us exactly to make this choice.

The equations of the base space are invariant under translations $a_{ij} \mapsto a_{ij} + \delta_i$, for all j; this is the action of the Lie algebra $\mathfrak{t}$ of diagonal matrices in $\mathfrak{gl}(n, \mathbb{C})$, induced by the action $x_i \mapsto x_i - \delta_i$ on $\mathbb{C}^n$. By taking a transverse slice to this action (e.g. the one given by $\sum_j a_{ij} = 0$) we obtain the miniversal deformation. This deformation space has been determined by various authors [Al, FP]. Our equations of are those of D. S. Rim (see [Scha]).

The structure of the base space has been studied in [St5]. It is not known if it is reduced for general n, but this is highly probable. The reduced base space S_{red} is the cone over a projective variety V_n, which is birational to $\mathbb{P}^{n-1} \times \mathbb{P}^{n-3}$; in particular, V_5 has as small resolution $\mathbb{P}^4 \times P$ with P a Del Pezzo surface of degree 5. The degree of V_n is determined in [GL]. For $n = 4$ the result is well known:

Lemma. *The base space of the miniversal deformation of L_4^4 is isomorphic to the cone over the Segre embedding of $\mathbb{P}^1 \times \mathbb{P}^3$.*

Proof. Each equation for the base space can be written as 2×2 determinant. We note the determinantal identity

$$\begin{vmatrix} a_{13} - a_{12} & a_{24} - a_{21} \\ a_{14} - a_{12} & a_{23} - a_{21} \end{vmatrix} = - \begin{vmatrix} a_{12} - a_{13} & a_{21} - a_{24} \\ a_{14} - a_{13} & a_{23} - a_{24} \end{vmatrix}.$$

Therefore we can write the equations as (2×2)-minors of

$$\begin{vmatrix} a_{12} - a_{13} & a_{21} - a_{24} & a_{34} - a_{31} & a_{43} - a_{42} \\ a_{14} - a_{13} & a_{23} - a_{24} & a_{32} - a_{31} & a_{41} - a_{42} \end{vmatrix}.$$

To get a base of minimal dimension, it is convenient to take $a_{13} = a_{24} = a_{31} = a_{42} = 0$, so we have indeed the required Segre cone.   □

Cotangent cohomology. We have given ad hoc definitions of T^1 and T^2. The theory of the cotangent complex enables one to define (in a functorial way) modules T^i for all $i \geq 0$. A down to earth construction is described in [Pa3]. Instead of a minimal free resolution of the analytic algebra $\mathcal{O}_X$ it uses the *Tyurina resolution*, which has in general infinite length. It has the structure of a free graded commutative algebra. Let a and b be homogeneous elements. Then

$$ba = -(-1)^{\deg a \, \deg b} ab.$$

We write as before $\mathcal{O}_X = \mathcal{O}_N/I$ and put $R_0 := \mathcal{O}_N$. A Tyurina resolution R is a free graded commutative R_0-algebra, generated by a set E of generators of negative degree, with finitely many generators in each degree. It has a differential s which is a derivation of degree 1, i.e.

$$s(ab) = s(a)b + (-1)^{\deg a} a s(b),$$

such that the sequence

$$0 \longleftarrow \mathcal{O}_X \longleftarrow R_0 \overset{s}{\longleftarrow} R_{-1} \overset{s}{\longleftarrow} R_{-2} \overset{s}{\longleftarrow} \ldots$$

is exact, where R_{-k} denotes the homogeneous part of degree $-k$. We construct it inductively. Let $(f_1, \ldots, f_{r_1})$ be a system of generators of the ideal I. We take R_{-1} the free R_0-module generated by generators $(e_1, \ldots, e_{r_1})$ and set $s(e_i) = f_i$. By the product rule s extends to a derivation on the free graded algebra generated by $(e_1, \ldots, e_{r_1})$. We obtain in this way the Koszul complex on the f_i. If X is a complete intersection and the F_i form a minimal set of generators we have a resolution of $\mathcal{O}_X$. If X is not a complete intersection, there are other relations between the f_i and we have to add generators $(e'_1, \ldots, e'_{r_2})$ of degree -2, and define the differential s on them in such a way that the $s(e'_i)$ span the kernel of $s\colon R_{-1} \to R_0$. As the e'_i commute we find that R has elements in all negative degrees. The differential extends to the free graded algebra generated by the e_i and the e'_j. Now we have to add generators in degree -3 and so on.

A derivation of degree k in the Tyurina resolution R is a $\mathbb{C}$-linear map $u\colon R \to R$ raising the grading by k and satisfying

$$u(ab) = u(a)b + (-1)^{k \deg a} a u(b)$$

on homogeneous elements. Together these derivations form a graded vector space $\mathrm{Der}(R) = \bigoplus_{-\infty}^{\infty} \mathrm{Der}_k(R)$. The bilinear operation

$$[u, v] = u \circ v - (-1)^{\deg u \deg v} v \circ u$$

makes it into a graded Lie algebra. The graded Jacobi identity reads

$$[u, [v, w]] = [[u, v], w] + (-1)^{\deg u \deg v} [v, [u, w]] .$$

The operator $d\colon u \mapsto [u, s]$ of degree 1 is a differential since $0 = 2s^2 = [s, s]$ and by the Jacobi identity

$$2d^2 u = 2[[u, s], s] = [u, [s, s]] = 0 .$$

Definition. The *cotangent cohomology* of the germ X is the cohomology of the complex $(\mathrm{Der}(R), d)$, denoted by $T^*(X)$.

The bracket operation descends to $T^*(X)$ and makes it into a graded Lie algebra. For other purposes it may be convenient to obtain $T^*(X)$ from a slightly different complex, see [Pa1]. We now show that the newly defined T^i coincide with the old ones.

Lemma. *Let $k > 0$. A class $[u] \in T^k(X)$ is completely determined by its values modulo I on the set of generators E_{-k} of R of degree $-k$. Every R_0-linear map $\bar{u}\colon R_{-k} \to \mathcal{O}_X$, vanishing on products of generators of degree greater than $-k$ and on $s(R_{-k-1})$ extends to an element of $\mathrm{Der}_k(R)$.*

Proof. Let u be a cocycle, i.e. $us - \pm su = 0$ with $u(e_i) \in I$ for all generators e_i of degree $-k$. We construct a v with $u = dv = vs - \mp sv$ by induction on the degree of the generators. As $u(e_i) \in I$ we can write $u(e_i) = s(a_i)$ for some $a_i \in R_{-1}$. We set $v(e_i) = -\mp a_i$ and zero on generators of degree $-l$ with $l < k$. Then $(vs - \mp sv)(e_i) = s(a_i) = u(e_i)$, establishing the base of the induction. Let now f_j be a generator of degree $-m$ and suppose $u = vs - \mp sv$ holds on all R_{-l} with $l < m$. Then $0 = (us - \pm su)(f_j) = (vs^2 - \mp svs - \pm su)(f_j)$. As $s^2 = 0$ we get $s(u(f_j) - vs(f_j)) = 0$ and by exactness of the Tyurina resolution $(u - vs)(f_j) = s(b_j)$ for some $b_j \in R_{-m-1}$ and we are done if we set $v(f_j) = -\mp b_j$.

Let $\bar{u}: R_{-k} \to \mathcal{O}_X$ be given. We define a derivation $u: R \to R$ by induction on the degree of the generators. On generators of degree greater than $-k$ we set u equal to zero. For $e_i \in E_{-k}$ we choose $u(e_i) \in R_0$ such that its class in $\mathcal{O}_X$ coincides with $\bar{u}(e_i)$. Let $f_j \in E_{-k-1}$. By assumption $us(f_j) = \pm s(b_j)$ for some $b_j \in R_{-1}$ and we set $u(f_j) = b_j$. Then $us - \pm su = 0$ on R_{-l} with $l \le k + 1$. If u is already defined on all R_{-l} with $l < m$ we get for a generator g_j of degree $-m$ that $su(s(g_j) = \pm us^2(g_j) = 0$ and as before exactness allows us to define $u(g_j)$ in the required manner. $\qquad\square$

By exactly the same arguments we conclude that $T^*(X)$ is concentrated in positive degrees.

Proposition. *One has $T^0(X) = \mathrm{Der}(\mathcal{O}_X)$ and one has the exact sequence*

$$0 \longrightarrow T^0(X) \longrightarrow (\mathrm{Der}\, R_0) \otimes \mathcal{O}_X \longrightarrow \mathrm{Hom}_{R_0}(I, \mathcal{O}_X) \longrightarrow T^1(X) \longrightarrow 0 \,.$$

Furthermore also $T^2(X)$ equals the earlier defined T_X^2.

Proof. An element $u \in T^0(X)$ defines a derivation of R_0 which satisfies $u(f_i) \in I$ for all generators f_i of I, as $us - su = 0$. Conversely, given a derivation of $\mathcal{O}_X$, we extend it to an element of $\mathrm{Der}_0(R)$ with the same construction as in the Lemma. This gives the first part of the exact sequence. From the Lemma we obtain the existence and surjectivity of the map $\mathrm{Hom}_{R_0}(I, \mathcal{O}_X) \to T^1(X)$. Its kernel consists of the image of $\mathrm{Der}\, R_0$. The statement about T^2 is checked in the same way. $\qquad\square$

Let R be a Tyurina resolution of the analytic algebra $\mathcal{O}_X$. We now perturb the differential s to a new differential s'. The cohomology of the complex (R, s') is again concentrated in degree zero and gives a deformation $\mathcal{O}_{X'}$ of our algebra, and all deformations can be obtained in this way. The condition that s' is a differential can be written as $[s', s'] = 0$. We are interested in 'small' deformations of s, so we write $s' = s + \varphi$. Then $[s', s'] = [s + \varphi, s + \varphi] = [s, s] + [s, \varphi] + [\varphi, s] + [\varphi, \varphi] = 2d\varphi + [\varphi, \varphi]$ and we get the deformation equation

$$d\varphi + \tfrac{1}{2}[\varphi, \varphi] = 0 \,.$$

Later on we will discuss strategies to solve the deformation equation. As defined, a perturbation φ is an infinite collection of maps. For practical purposes it is better to represent deformations by a finite collection of maps. But that is precisely what we did when we required exactness in the minimal free resolution and considered the equation $FR = 0$.

The cotangent cohomology described is part of an even larger theory. We constructed the Tyurina resolution R of the algebra $\mathcal{O}_X$ over the basering $\mathbb{C}$, but one can work over any basering. So let A be an analytic algebra and let the analytic algebra B be an A-algebra and construct a Tyurina resolution $R_{B/A}$. For any B-module M we can now define

$$T^*(B/A, M) = H^*(\mathrm{Der}(R_{B/A}) \otimes_B M) \ .$$

The complex $\mathrm{Der}(R_{B/A})$ is an incarnation of the *tangent complex*, which is the dual of the *cotangent complex*, denoted by $\mathbb{L}_{B/A}$. A direct definition of the cotangent complex from the Tyurina resolution is given in [Pa1, Pa2]. One can also write $T^*(B/A, M) = H^*(\mathrm{Hom}_B(\mathbb{L}_{B/A}, M))$. Note that $T^*(X) = T^*(\mathcal{O}_X/\mathbb{C}, \mathcal{O}_X)$.

We conclude by listing some properies of cotangent cohomology. Proofs can be found in [Pa2].

1. A short exact sequence $0 \to M' \to M \to M'' \to 0$ of B-modules induces a long exact sequence

$$0 \to T^0(B/A, M') \longrightarrow T^0(B/A, M) \longrightarrow$$
$$\longrightarrow T^0(B/A, M'') \longrightarrow T^1(B/A, M') \longrightarrow \cdots \ .$$

2. One has

$$T^0(B/A, M) = \mathrm{Hom}(\Omega_{B/A}, M) = \mathrm{Der}_A(B, M) \ .$$

3. If B is a regular A-algebra, then $T^i(B/A, M) = 0$ for all $i > 0$.
4. *Base change.* If B is a flat A-module and B' is obtained by base-change from a map $A \to A'$ and M' is a B' module, then there is an isomorphism

$$T^i(B'/A', M') \cong T^i(B/A, M') \ .$$

5. *Zariski-Jacobi sequence.* Let $A \to B \to C$ be morphisms of analytic algebras and M a C-module. There is a long exact sequence

$$\cdots \longrightarrow T^i(C/B, M) \longrightarrow T^i(C/A, M) \longrightarrow$$
$$\longrightarrow T^i(B/A, M) \longrightarrow T^{i+1}(C/B, M) \longrightarrow \cdots \ .$$

4 Example: the fat point of multiplicity four

In this section we give another example of the computation of a versal deformation: we describe it for the fat point Z_3 of multiplicity 4, with local ring $\mathcal{O}_3/\mathfrak{m}_3^2$. We take coordinates x_1, x_2 and x_3 on $\mathbb{C}^3$. We will frequently use summation over indices, running from 1 to 3; as we do not write this range explicitly, we might as well pretend, that the summation runs from 1 to n, and do the computation for the fat point Z_n, with local ring $\mathcal{O}_n/\mathfrak{m}_n^2$. In the end this maybe simplifies matters, because it stops you taking 'shortcuts', which use the fact that there are only three indices, but obscure the structure. The base space of Z_n has been obtained around 1980 by Frank SCHREYER, by André GALLIGO (although he made some mistakes in the simplification of the equations in the case of Z_3) and by Ragnar BUCHWEITZ. The minimal system of generators for the ideal of the base seems to be new.

We take coordinates $x_1, \ldots, x_n$ on $\mathbb{C}^n$. The ideal I_n of Z_n is generated by all $\binom{n+1}{2}$ quadratic monomials in the x_i. Obviously the full symmetric group S_n acts; even more is true, the ideal I_n is Gl_n-invariant, and by general theory the versal deformation can be chosen Gl_n-equivariant [Rim2]. We will use the S_n-action only on a naive level, in that we assume that the set of all our equations is invariant under permutation of the indices. The index i in x_i is a kind of free variable, which can take any value from 1 to n. We note that there are two types of equation: $x_i x_j$ with $i \neq j$, and x_i^2; if we allow in $x_i x_j$ also the possibility $i = j$, then we can write all equations in this way. The relations are

$$(x_i x_j) x_k - (x_i x_k) x_j . \tag{4.1}$$

Consider infinitesimal deformations $x_i x_j + \phi_{ij}$; from the relations we get that $\phi_{ij} \in \mathfrak{m}$, but besides that there are no conditions, so $\dim T^1 = n\binom{n+1}{2} - n$ (subtract n dimensions coming from coordinate transformations). We take coordinates a_{ij}^k on T^1, and write as deformed equations

$$x_i x_j + \sum_k a_{ij}^k x_k .$$

It is understood that $a_{ij}^k = a_{ji}^k$. To get a basis for T^1 we may take $a_{ii}^i = 0$, but to avoid different cases of the equations, we work with the slightly larger number of $n\binom{n+1}{2}$ variables, or in other words, with a basis of $H^0(Z_n, N)$.

The obstruction calculus will give some conditions on the a_{ij}^k, and we will find for each equation a term b_{ij}, which is quadratic in the a_{ij}^k, well determined up to the equations between them. This gives then

$$x_i x_j + \sum_k a_{ij}^k x_k + b_{ij} \tag{4.2}$$

as final form of the equations. We take the b_{ij} as extra variables. Alternatively, the equations (4.2) can be seen as unfolding of the original ones, and we are asking for the *flattener*; this is the method of [Gal].

We lift the relations (4.1):

$$\left(x_i x_j + \sum_l a_{ij}^l x_l + b_{ij}\right) x_k - \left(x_i x_k + \sum_l a_{ik}^l x_l + b_{ik}\right) x_j$$
$$-\sum_l \left(x_l x_k + \sum_m a_{lk}^m x_m + b_{lk}\right) a_{ij}^l + \sum_l \left(x_l x_j + \sum_m a_{lj}^m x_m + b_{lj}\right) a_{ik}^l =$$
$$b_{ij} x_k - b_{ik} x_j - \sum_{l,m} a_{ij}^l a_{lk}^m x_m + \sum_{l,m} a_{ik}^l a_{lj}^m x_m - \sum_l a_{ij}^l b_{lk} + \sum_l a_{ik}^l b_{lj} \,.$$

The right hand side of this equation is at least quadratic in the a_{ij}^k (given our assumption on the b_{ij}). To have a lift of the relations it has to vanish. The coefficients of the x_i give the following equations:

$$b_{ij} - \sum_l a_{ij}^l a_{lk}^k + \sum_l a_{ik}^l a_{lj}^k = 0 \,, \qquad k \neq j \,, \tag{4.3}$$

and

$$\sum_l a_{ij}^l a_{lk}^m - \sum_l a_{ik}^l a_{lj}^m = 0 \,, \qquad m \neq j, k \,. \tag{4.4}$$

We introduce for all values of the indices the abbreviation

$$F_{i,jk}^m = \sum_l a_{ij}^l a_{lk}^m - \sum_l a_{ik}^l a_{lj}^m \,.$$

The equations (4.4) are $F_{i,jk}^m = 0$ for $m \neq j, k$, and (4.3) becomes $b_{ij} - F_{i,jk}^k = 0$. We remark that this formula is not symmetric in i and j, but we note the following identities:

$$\begin{aligned} F_{i,jk}^m + F_{i,kj}^m &= 0 \,, \\ F_{i,jk}^m + F_{j,ki}^m + F_{k,ij}^m &= 0 \,, \\ \sum_{l \neq i,j} F_{l,ij}^l + F_{j,ij}^j - F_{i,ji}^i &= 0 \,. \end{aligned} \tag{4.5}$$

The last identity follows from

$$\sum_{k,l} a_{li}^k a_{kj}^l - \sum_{k,l} a_{lj}^k a_{ki}^l = 0 \,.$$

That in the lift of the relations also the term $R = \sum a_{ij}^l b_{lk} - \sum a_{ik}^l b_{lj}$ vanishes follows from the fact that T^2 is concentrated in degree -2, but it can also be seen by direct computation; we first use (4.3) with i as repeated index to eliminate the b_{kl} and b_{jl}, which works for $l \neq i$, so

$$R = a_{ij}^i b_{ki} - a_{ik}^i b_{ji} + {\sum}' a_{ij}^l a_{kl}^m a_{mi}^i - {\sum}' a_{ij}^l a_{ki}^m a_{ml}^i$$
$$- {\sum}' a_{ik}^l a_{jl}^m a_{mi}^i + {\sum}' a_{ik}^l a_{ji}^m a_{ml}^i , \qquad (4.6)$$

where $\sum'$ stands for summation over m and over $l \neq i$. The terms of the second and fourth sum are the same, except that l and m are interchanged, and therefore the terms with $m \neq i$ cancel, leaving

$$\sum_{l \neq i} a_{il}^i (a_{ij}^i a_{ik}^l - a_{ij}^l a_{ki}^i) = \sum_l a_{il}^i (a_{ij}^i a_{ik}^l - a_{ij}^l a_{ki}^i) .$$

If we replace the index l by m, we recognise this term as the missing $l = i$ case of the first and third sum in (4.6). Therefore

$$R = a_{ij}^i b_{ki} - a_{ik}^i b_{ji} + \sum_{l,m} a_{mi}^i (a_{ij}^l a_{lk}^m - a_{ik}^l a_{lj}^m) .$$

The term in parentheses vanishes by (4.4) for $m \neq j, k$; for these two values we have a single summation over l. Collecting terms gives

$$R = a_{ij}^i (b_{ik} - F_{i,kj}^j) - a_{ik}^i (b_{ij} - F_{i,jk}^k) = 0 .$$

Theorem. *A minimal set of generators of the ideal of the base space of Z_n consists of*

$$\begin{aligned}
&\text{for } i < j < k, \ m \neq i,j,k: &&F_{i,jk}^m , &&F_{j,ki}^m , &&F_{i,jk}^i , &&F_{j,ki}^j , &&F_{k,ij}^k , \\
&\text{for } i < j, \ m \neq i,j: &&F_{i,ji}^i - F_{i,jm}^m , &&F_{i,ij}^m , &&F_{j,ij}^m , \\
&\text{for } i, \ \text{fix a } j \neq i, \ m \neq i,j: &&F_{i,ij}^j - F_{i,im}^m .
\end{aligned}$$

These equations are to be considered as involving $n\binom{n+1}{2} - n$ variables, i.e., we set $a_{ii}^i = 0$. Without this condition the equations in $n\binom{n+1}{2}$ variables describe the Hilbert scheme $\mathrm{Hilb}^{n+1} \mathbb{C}^n$. The number of equations is $2n\binom{n+1}{3} - \binom{n+1}{2} - \binom{n}{2}$, whereas the dimension of T^2 is $2n\binom{n+1}{3} - \binom{n+1}{2}$.

Proof. We obtain the equations from (4.3) and (4.4) by eliminating the b_{ij}. The given set of equations generate, in view of the relations (4.5). We have to show that these quadratic equations are linearly independent. For this purpose it suffices to exhibit in each equation a monomial occurring only in this equation. A term $a_{\cdots}^m a_{\cdots}^m$ comes from an expression $F_{\cdots}^m$. In $F_{i,jk}^m$ we have $a_{ij}^m a_{mk}^m - a_{ik}^m a_{mj}^m$, which cancels out only for $m = i$. Both monomials determine the set $\{i, j, k, m\}$. For $i < j < k$ we have $F_{i,jk}^m$ and $F_{j,ki}^m$; in the last equation the terms are $a_{jk}^m a_{mi}^m - a_{ij}^m a_{mk}^m$, so the monomial $a_{ik}^m a_{mj}^m$ occurs exclusively in $F_{i,jk}^m$, and $a_{jk}^m a_{mj}^m$ in $F_{j,ki}^m$. Writing out the terms for the remaining equations with m as upper index shows $a_{im}^m a_{jm}^m$ to occur only in $F_{i,jm}^m$, and $a_{ij}^m a_{mi}^m$ in $F_{i,ij}^m$, for arbitrary i, j. Finally, we have to show that the $F_{i,jk}^i$ are linearly independent. We look only at the monomials which involve three indices:

$$a_{ij}^j a_{jk}^i - a_{ik}^j a_{jj}^i + a_{ij}^k a_{kk}^i - a_{ik}^k a_{jk}^i .$$

In each of these monomials the index i is the only one occurring twice, so they appear only in the equation $F_{i,jk}^i$.

The number of equations is $(2n-3)\binom{n}{3} + 3(n-2)\binom{n}{2} + n(n-2)$. As to T^2, the $\mathcal{O}_{Z_n}$-module $\mathcal{R}$ of relations is a $\mathbb{C}$-vector space of dimension $2\binom{n+1}{3}$ with trivial module structure, and $\operatorname{Hom}_{Z_n}(\mathcal{R}, \mathcal{O}_{Z_n}) = \operatorname{Hom}_{\mathbb{C}}(\mathcal{R}, \mathfrak{m}/\mathfrak{m}^2)$: every function on relations maps into the maximal ideal, and apart from that there are no conditions. The equations of Z_n span a $\binom{n+1}{2}$-dimensional subspace. $\square$

We analyse the structure of the base space only in the case $n = 3$; for $n = 4$ we have already 64 equations in 36 variables, and the base space is highly singular. In fact, the deformation theory of the spaces Z_n contains the deformation theory of all fat points of multiplicity at most $n+1$.

Proposition. *Every fat point Y of multiplicity $n+1$ is a deformation of Z_n.*

Example. Let the ideal of Y be (x^2, y^2). Introduce a new variable z, such that Y is given by (z, x^2, y^2), and make a change of variables: replace z by $z - xy$. A non minimal set of generators is $(z - xy, x^2, y^2, zx, zy, z^2)$. Now homogenise to $(tz - xy, x^2, y^2, zx, zy, z^2)$. This realises the original ideal as deformation of Z_3.

Proof of the Proposition. Suppose Y is minimally embedded in $\mathbb{C}^m$ with $m < n$. Choose a monomial basis $e_{m+1}, \dots, e_n$ of $\mathfrak{m}_Y^2$, and take extra variables $x_{m+1}, \dots, x_n$, and add equations $x_i - e_i = 0$. From the multiplication in $\mathcal{O}_Y$ we obtain constants a_{ij}^k such that in $\mathcal{O}_Y$

$$x_i x_j + \sum_k a_{ij}^k x_k = 0 \, .$$

These equations generate the ideal of Y in $\mathbb{C}^n$. Now homogenise: replace a_{ij}^k by $t a_{ij}^k$. For $t \neq 0$ we have a singularity, isomorphic to Y, and for $t = 0$ we have Z_n; the family is flat, because the multiplicity is constant. $\square$

The group of affine transformations acts on $\operatorname{Hilb}^{n+1} \mathbb{C}^n$, and it has a dense orbit in the component parametrising $n+1$ distinct points. This follows from the following well known fact:

Proposition. *Two sets of $n+1$ points in affine n-space in general position are affinely equivalent.*

Therefore Gl_n acts on the base space of Z_n with a dense orbit in the smoothing component. Because not all fat points are smoothable the base space is in general reducible.

We now come to the deformation space of Z_3. There are 15 equations for the base. We write five of them, the other ten are obtained by cyclic permutation of the indices:

$$F_{1,23}^1 : a_{13}^2 a_{22}^1 - a_{23}^1 (a_{12}^2 - a_{13}^3) - a_{12}^3 a_{33}^1$$
$$F_{1,31}^2 : a_{13}^2 (a_{12}^2 + a_{13}^3 - a_{11}^1) + a_{11}^2 (a_{31}^1 - a_{32}^2) - a_{11}^3 a_{33}^2$$
$$F_{1,12}^3 : -a_{12}^3 (a_{12}^2 + a_{13}^3 - a_{11}^1) + a_{11}^3 (a_{23}^3 - a_{21}^1) + a_{11}^2 a_{22}^3$$
$$F_{1,13}^3 - F_{1,12}^2 :$$
$$(a_{12}^2 - a_{13}^3)(a_{12}^2 + a_{13}^3 - a_{11}^1) + a_{11}^2 (a_{23}^3 + a_{21}^1 - a_{22}^2) - a_{11}^3 (a_{23}^3 + a_{21}^1 - a_{33}^3)$$
$$F_{1,31}^1 - F_{3,12}^2 : -(a_{12}^2 - a_{13}^3)(a_{31}^1 - a_{32}^2) + a_{13}^2 (a_{23}^3 + a_{21}^1 - a_{22}^2) - a_{11}^3 a_{33}^1 \ .$$

The parentheses in these equations suggest a coordinate transformation, from which it can be seen that the miniversal base space of Z_3 is the cone over the Plücker embedding of the Graßmannian G_2^6, an observation which I learned from Jan CHRISTOPHERSEN. Sheldon KATZ gives a different explanation, which identifies a point in the projectivised base space with a pencil of conics in $\mathbb{P}^2(\mathbb{C})^*$ [Ka]. Let V be the six dimensional vector space of plane conics. The group PGl_3 acts on the Graßmannian $G_2^6 = G_2(V)$, and it has a dense open orbit. As we have seen, the action of Gl_3 on the base space has also a dense orbit. Therefore it suffices to give a Gl_3 equivariant isomorphism of $\bigwedge^2 V$ with T^1, which identifies a general one parameter smoothing with a point of the Graßmannian.

A very simple smoothing is the determinantal deformation

$$\text{Rank} \begin{pmatrix} t & x_1 & x_2 & x_3 \\ x_1 & x_2 & x_3 & t \end{pmatrix} \leq 1 \ .$$

The equations are not in the form (4.2), but they can be easily rewritten. For a fixed $t \neq 0$ we have the four points $p_k = (i^k t, i^{2k} t, i^{3k} t)$, where $i = \sqrt{-1}$ and $k = 0, \ldots, 3$. We remark that $p_1 + p_2 + p_3 + p_4 = 0$, so the centre of gravity is the origin.

We will use the remaining freedom in the choice of the deformation parameters, to place the centre of gravity always at the origin. To find the centre of gravity, we eliminate in the equations (4.2) two space variables, say x_1 and x_2, to obtain a quartic monic polynomial in x_3, and the coefficient of x_3^3 is the x_3-coordinate of the centre of gravity. In this way we find three equations:

$$a_{13}^1 + a_{23}^2 + a_{33}^3 = 0$$
$$a_{12}^1 + a_{22}^2 + a_{23}^3 = 0$$
$$a_{11}^1 + a_{12}^2 + a_{13}^3 = 0 \ .$$

We can associate a point of $G_2(V)$ to a general one parameter smoothing by the following construction. However, it does not give the identification of the projectivised base space with the Graßmannian, but apparently it leads to a birational automorphism.

Construction. Consider four points $p_1, \ldots, p_4$ in $\mathbb{C}^3$ in general position, with centre of gravity at the origin. By assumption the points are linearly independent, and therefore distinct from the origin. Let $\pi: \mathbb{C}^3 \setminus \{0\} \to \mathbb{P}^2$ be the projection, then the four points $\pi(p_1), \ldots, \pi(p_4)$ in general position in $\mathbb{P}^2$ determine a pencil of conics. These conics can be computed from the ideal of the points $p_1, \ldots, p_4$, which is generated by equations (4.2) for some constants a_{ij}^k, b_{ij}. Homogenising these equations with respect to a variable t realises the points as deformation of Z_3. Eliminating t from the homogenised equations gives two quadratic equations in x_1, x_2 and x_3.

Conversely, a general pencil of conics has four base points in $\mathbb{P}^2$, and determines four lines L_i through the origin in $\mathbb{C}^3$. Choose a point p_i on each line L_i, such that $p_1 + p_2 + p_3 + p_4 = 0$. This condition gives three linear homogeneous equations in four variables, with one dimensional solution space: the pencil determines a line in the base space of Z_3.

It is tempting to ask for a 'natural' identification of the projectivised base space with $G_2(V)$. But maybe there is no deeper explanation than the fact that the Graßmannian is the only Gl_3-manifold of the correct dimension and embedding dimension.

5 Deformations of algebras

Deformations of singularities can be described in terms of defining equations but also in terms of change of the multiplication law of a commutative ring. In this section we start with an elementary treatment of deformations of associative algebras, following [Nij].

Let A be an algebra over the field k. This means that A has the underlying structure of a k-vector space and in addition elements of A can be multiplied. The multiplication map $\mu\colon A \times A \to A$ is k-bilinear and we write $\mu(a, b)$ or more simply ab. In case A is finite dimensional the multiplication is completely described by a set of structure constants c_{ij}^k: let $(e_1, \ldots, e_n)$ be a basis of A then

$$\mu(e_i, e_j) = \sum_k c_{ij}^k e_k \ .$$

Until now we have said nothing about associativity or any other property of the multiplication μ. Important classes of algebras are equationally defined: commutative ones by $ab = ba$, associative ones by $a(bc) = (ab)c$. The last condition reads in terms of μ:

$$\mu\big(a, \mu(b, c)\big) - \mu\big(\mu(a, b), c\big) = 0 \ .$$

Another example are Lie algebras: the product is anti-symmetric and satisfies the Jacobi identity. Traditionally the product is written with brackets $[a, b]$, whereas purists insist on ab — the way the product in an algebra without any additional structure is written. We will reserve the bracket notation for commutators: $[a, b] = ab - ba$, and use μ for the product.

Now we want to deform the algebra A. We do this by changing the structure constants. So by definition the underlying vector space structure remains the same, but we change the multiplication μ. We start by looking at 1-parameter deformations and we write μ_t. Our task is to describe the possible perturbations (up to isomorphism). To make progress we have to choose a class of algebras. From now on we consider finite dimensional *associative* algebras.

We want to describe infinitesimal deformations of A. We start with a heuristic derivation of the relevant equation. Consider an arbitrary family μ_t of associative algebra structures with $\mu_0 = \mu$. So we have

$$\mu_t\big(a, \mu_t(b, c)\big) - \mu_t\big(\mu_t(a, b), c\big) = 0 \ .$$

Let's further assume that μ_t depends differentiably on t, meaning that the structure constants are differentiable functions of t. This presupposes of course that $k = \mathbb{R}$ or $k = \mathbb{C}$. Write f for $\frac{d}{dt}\mu_t|_{t=0}$. Then by the product rule

$$f\big(a, \mu(b,c)\big) + \mu\big(a, f(b,c)\big) - f\big(\mu(a,b), c\big) - \mu\big(f(a,b), c\big) = 0 \,,$$

which we can write shorter as

$$f(a, bc) + af(b,c) - f(ab, c) - f(a,b)c = 0 \,.$$

The solutions of this equation are by definition the first order infinitesimal deformations.

Now suppose we have given a k-linear map $f: A \to A$. Then e^{tf} is an invertible k-linear map and we get a family of isomorphic algebra structures by defining

$$\mu_t(a, b) = e^{-tf}\mu(e^{tf}a, e^{tf}b) \,.$$

Again we derive and find

$$\frac{d}{dt}\mu_t|_{t=0}(a,b) = -f\mu(a,b) + \mu(fa, b) + \mu(a, fb) = -f(ab) + f(a)b + af(b) \,.$$

We can view the formulas above as the rule for the coboundary operator in a complex. The relevant theory is *Hochschild cohomology* (a good reference is [Lod]). Let M be an A-bimodule, so we have two bilinear operations $\lambda: A \times M \to M$ and $\rho: M \times A \to M$ satisfying $\lambda\big(a, \lambda(b, m)\big) = \lambda\big(\mu(a,b), m\big)$ or more shortly $a(bm) = (ab)m$ and similarly for the right module structure $(ma)b = m(ab)$; moreover $(am)b = a(mb)$ or $\rho\big(\lambda(a,m), b\big) = \lambda\big(a, \rho(m, b)\big)$. Clearly A itself is an A-bimodule (with $\lambda = \rho = \mu$).

Definition. The k-module of Hochschild n-cochains of A with values in M is the vector space $C^n(A; M)$ of k-multilinear functions of n variables $f: A \times \cdots \times A \to M$. The coboundary map $\delta: C^n(A; M) \to C^{n+1}(A; M)$ given by

$$\delta f(a_0, \ldots, a_n) = a_0 f(a_1, \ldots, a_n) - f(a_0 a_1, a_2, \ldots, a_n)$$
$$+ f(a_0, a_1 a_2, \ldots, a_n) + \cdots + (-1)^{n-1} f(a_0, \ldots, a_{n-2} a_{n-1}, a_n)$$
$$+ (-1)^n f(a_0, \ldots, a_{n-2}, a_{n-1} a_n) + (-1)^{n+1} f(a_0, \ldots, a_{n-1}) a_n$$

or

$$\delta f(a_0, \ldots, a_n) = \lambda\big(a_0, f(a_1, \ldots, a_n)\big)$$
$$+ \sum (-1)^i f(a_0, \ldots, \mu(a_{i-1}, a_i), \ldots, a_n) + (-1)^{n+1} \rho\big(f(a_0, \ldots, a_{n-1}), a_n\big) \,.$$

One can check directly that $\delta \circ \delta = 0$ (which we will not do here), so $C^\bullet$ is a complex. Its cohomology is Hochschild cohomology.

Let us take a closer look at low values of n. For $n = 0$ an element $f \in C^0(A; M)$ is just an $m \in M$ and

$$(\delta m)(a) = am - ma .$$

Now $(\delta m)(ab) = (ab)m - m(ab) = (am - ma)b + a(bm - mb) = (\delta m)(a)b + a(\delta m)(b)$ so δm is the 'interior derivation' of A into M determined by m.

If $n = 1$ we have a linear map $f \colon A \to M$. Suppose $\delta f = 0$, so

$$0 = (\delta f)(a, b) = af(b) - f(ab) + f(a)b$$

or

$$f(ab) = af(b) + f(a)b .$$

Therefore f is a derivation of A into M. In case $M = A$ we found for $\mu_t(a, b) = e^{-tf}\mu(e^{tf}a, e^{tf}b)$ that

$$\frac{d}{dt}\mu_t|_{t=0}(a, b) = (\delta f)(a, b) .$$

Similarly, for $M = A$ and $n = 2$ a multiplication $\mu_t = \mu + tf$ satisfies up to terms of second order the associativity constraint if and only if $\delta f = 0$. We conclude that the Hochschild cohomology $H^2(A; A)$ classifies infinitesimal deformations modulo isomorphism.

We give a different interpretation of deformations, in terms of extensions. Let M be again a A-bimodule and consider the semi-direct product $B = A \times M$ with multiplication $\overline{\mu} \colon B \times B \to B$ given by

$$\overline{\mu}\big((a, m), (b, n)\big) = \big(\mu(a, b), \lambda(a, n) + \rho(m, b)\big) .$$

Then B is again an associative algebra, which satisfies the following properties:

(i) M is an ideal with $M^2 = 0$,
(ii) the quotient B/M is isomorphic to A as algebra, and
(iii) M is a module over B/M via λ and ρ.

The *extension problem* asks for all exact sequences

$$0 \longrightarrow M \longrightarrow B \longrightarrow A \longrightarrow 0$$

satisfying (i), (ii) and (iii). The underlying vector space of each such a B is still $A \times M$ and we are asking for all multiplications $\overline{\mu}'$ on $A \times M$ satisfying (i), (ii) and (iii). Remark that A is not necessarily a subalgebra. From (i) we have that $\overline{\mu}'$ vanishes on M, by (ii) $\overline{\mu}$ and $\overline{\mu}'$ differ on A by a map φ with values in M while (iii) implies that they coincide when evaluated on an element of B/M and of M. So we can write:

$$\overline{\mu}'\big((a, m), (b, n)\big) = \big(\mu(a, b), \varphi(a, b) + \lambda(a, n) + \rho(m, b)\big) .$$

A direct computation shows that $\overline{\mu}'$ is associative if and only if

$$\lambda\big(a, \varphi(b, c)\big) - \rho\big(\varphi(a, b), c\big) + \varphi\big(a, \mu(b, c)\big) - \varphi\big(\mu(a, b), c\big) = 0 ,$$

or $\delta\varphi = 0$. Two extensions are equivalent if one has a commutative diagram

$$
\begin{array}{ccccccccc}
0 & \longrightarrow & M & \longrightarrow & B & \longrightarrow & A & \longrightarrow & 0 \\
& & \| & & \downarrow & & \| & & \\
0 & \longrightarrow & M & \longrightarrow & B' & \longrightarrow & A & \longrightarrow & 0 .
\end{array}
$$

The isomorphism in the middle is of the form $F(a, m) = \big(a, m + f(a)\big)$ with $f : A \to M$ a k-linear map, its inverse being $F^{-1}(a, m) = \big(a, m - f(a)\big)$. The structures which are equivalent to $\bar{\mu}$ are

$$
\begin{aligned}
\bar{\mu}'\big((a, m), (b, n)\big) &= F^{-1}\bar{\mu}\big(F(a, m), F(b, n)\big) \\
&= F^{-1}\big(\mu(a, b), \lambda(a, n + f(b)) + \rho(m + f(a), b)\big) \\
&= \big(\mu(a, b), -f\big(\mu(a, b)\big) + \lambda(a, n) + \lambda\big(a, f(b)\big) + \rho(m, b) + \rho(f(a), b)\big)
\end{aligned}
$$

and we find in this case that

$$
\varphi(a, b) = -f\big(\mu(a, b)\big) + \lambda\big(a, f(b)\big) + \rho(f(a), b) = (\delta f)(a, b) .
$$

We conclude that the Hochschild cohomology group $H^2(A; M)$ classifies extensions of A by M modulo equivalence. To complete the picture we recall the *Baer sum* of extensions. Given two such we first form the direct sum

$$
0 \longrightarrow M \oplus M \longrightarrow B \oplus B' \longrightarrow A \oplus A \longrightarrow 0
$$

then pull-back along the diagonal map $A \to A \oplus A : a \mapsto (a, a)$ and push-out along the addition map $M \oplus M \to M : (m, n) \mapsto m + n$: first let $\tilde{B}$ be the subalgebra of pairs (b, b') having the same image in A and then form the quotient $\hat{B} = \tilde{B}/\{(m, -m)\}$. Then

$$
0 \longrightarrow M \longrightarrow \hat{B} \longrightarrow A \longrightarrow 0
$$

is the required extension.

In particular if $M = A$ an extension may be regarded as a $k[t]/(t^2)$-algebra structure on the module $A[t]/(t^2)$ and $H^2(A; A)$ classifies $k[t]/(t^2)$-algebras which are isomorphic as modules to $A[t]/(t^2)$. This is a more precise or better statement then the previous one based on happily differentiating the structure constants. Given a k-algebra A, a deformation of A is an object over a base ring R, so we should indeed consider R-algebra structures on $A \otimes R$. One possibility for R is the ring $k[[t]]$ of formal power series in one variable. Then the structure constants c_{ij}^k are formal power series in t and the only value of t at which we can evaluate is $t = 0$. Of course if the constants are analytic functions of t we can evaluate at other points, and speak loosely about structures μ_t on A.

Our next goal is to describe higher order deformations. To this end we first introduce a new product. Consider a vector space V, at the moment without any further algebraic structure, and multilinear maps $V^{\oplus n} \to V$.

Definition. The *composition product* $g \bar{\circ} f$ of a n-linear map f and a m-linear map g is the $(n + m - 1)$-linear map given by

$$(g \bar{\circ} f)(x_1, \ldots, x_{n+m-1}) =$$
$$\sum_{i=1}^{n} (-1)^{(i-1)(n-1)} g(x_1, \ldots, x_{i-1}, f(x_i, \ldots, x_{i+n-1}), x_{i+n}, \ldots, x_{n+m-1}).$$

Example. Let $\mu: V \times V \to V$ be bilinear. Then

$$(\mu \bar{\circ} \mu)(x, y, z) = \mu\big(\mu(x, y), z\big) - \mu\big(x, \mu(y, z)\big)$$

and the condition $\mu \bar{\circ} \mu = 0$ is exactly that μ defines an associative multiplication on V.

Example. Let f be n-linear and μ an associative algebra structure on the algebra A. Then

$$\delta f = (-1)^{n+1} \mu \bar{\circ} f - f \bar{\circ} \mu.$$

We are now in a position to define on the Hochschild complex $C^{\bullet}(A, A)$ the structure of a graded Lie algebra. Note that infinitesimal deformations are parametrised by H^2 instead of a first cohomology group. Therefore we introduce a reduced degree $\bar{n} := n - 1$.

Theorem. . *Let V a vector space and $C^{\bullet}(V)$ be the complex of multilinear maps of V into itself. Then*

$$[f, g]^{\circ} = g \bar{\circ} f - (-1)^{\bar{n}\,\bar{m}} f \bar{\circ} g$$

defines a graded Lie algebra structure on $C^{\bullet}(V)$ with respect to reduced degree. In particular one has

$$[f, g]^{\circ} + (-1)^{\bar{n}\,\bar{m}} [g, f]^{\circ} = 0$$

and the graded Jacobi identity

$$(-1)^{\bar{n}\,\bar{p}} \big[[f, g]^{\circ}, h\big]^{\circ} + (-1)^{\bar{m}\,\bar{n}} \big[[g, h]^{\circ}, f\big]^{\circ} + (-1)^{\bar{p}\,\bar{m}} \big[[h, f]^{\circ}, g\big]^{\circ} = 0.$$

'*Proof*'. For hints fto the proof we refer to [Nij]. One can start by proving the following identity for maps f, g and h which take n, m and p arguments:

$$(f \bar{\circ} g) \bar{\circ} h - f \bar{\circ} (g \bar{\circ} h) = (-1)^{\bar{m}\,\bar{p}} \big((f \bar{\circ} h) \bar{\circ} g - f \bar{\circ} (h \bar{\circ} g)\big).$$

For this one needs 'patience, a big sheet of paper, a sharp pencil and good light'.   □

From the second example above we have that for an associative multiplication

$$\delta f = -[\mu, f]^{\circ}.$$

As the reduced degree of μ is one, we find

$$[\mu, \mu]^\circ = 2\,\mu\bar{o}\mu \,,$$

which vanishes if μ is associative. The Jacobi identity yields

$$\delta^2 f = \big[\mu, [\mu, f]^\circ\big]^\circ = \tfrac{1}{2}\big[[\mu, \mu]^\circ, f\big]^\circ = 0 \,.$$

The proof that $\delta^2 = 0$ on $C^\bullet(A; M)$ can be reduced to this special case by means of the semi-direct product $B = A \times M$. In a natural way $C^\bullet(A; M)$ is a sub-complex of $C^\bullet(B; B)$. Also

$$\delta[f, g]^\circ = [\delta f, g]^\circ + (-1)^{\bar{n}}[f, \delta g]^\circ \,.$$

This formula shows that the bracket $[-, -]^\circ$ descends to a Lie algebra structure on the cohomology ring $H^\bullet(A; A)$.

Now we return to deformations. Suppose we have an algebra A with an associative multiplication μ, so $\mu\bar{o}\mu = 0$. We are interested in 'small' deformations of μ. We write $\mu + \varphi$ for a deformed multiplication with φ bilinear. The associativity condition is

$$0 = (\mu + \varphi)\bar{o}(\mu + \varphi) = \mu\bar{o}\mu + \mu\bar{o}\varphi + \varphi\bar{o}\mu + \varphi\bar{o}\varphi = -\delta\varphi + \tfrac{1}{2}[\varphi, \varphi]^\circ$$

so we have the *deformation equation*

$$\delta\varphi - \tfrac{1}{2}[\varphi, \varphi]^\circ = 0 \,.$$

We describe methods to solve it in Chap. 8.

Remark. If A is commutative we want to consider deformations which are also commutative. We can work with the Harrison complex, which is a sub-complex of the Hochschild complex. For $C^2(A; M)$ we look only at symmetric f satisfying $f(a, b) = f(b, a)$. For $n = 0$ and $n = 1$ Hochschild and Harrison cochains coincide, for $n > 2$ the definition is a bit complicated (see [Lod]).

If A is an analytic algebra some of the definitions have to be slightly adapted to take convergence into account. Details are in a paper by PALAMODOV [Pa2]. In this case one can also compute with the *Tyurina resolvent*. In the quoted paper PALAMODOV proves that the resulting cohomology is the same by constructing a double complex combining both constructions. This is an instance of the phenomenon that for theoretical purposes big complexes are useful, whereas for computations one wants small complexes. I do not know any example of a computation of deformation spaces using Harrison cohomology.

The deformation theory of Lie algebras uses analogously Lie algebra cohomology.

6 Formal deformation theory

In a seminal paper SCHLESSINGER gave conditions for the existence of semi-universal deformations as formal objects [Schl1]. The disadvantage of them is that one gets a rather weak result; on the other hand, it applies to a large class of deformation problems. The general theory uses the language of categories, so we have to immerse in some abstract definitions. Our basic reference is the SGA-exposé [Rim1]; the set-up is more general than in [Schl1], but actually the treatment is closer to our previous discussion of the main example, deformations of singularities.

Deformations of germs form a category, which we now describe in terms of local rings instead of spaces. All algebras in this paragraph will be analytic algebras, objects in the category (Analg). An object in our deformation category Def is a homomorphism $R \to A$, which makes A into a flat R-algebra. A morphism $A \to B$ over $R \to S$ is an S-algebra isomorphism $\phi: A \otimes_R S$. Given a morphism $R \to S$ and an object A, we have a lift $A \to A \otimes_R S$. Furthermore, given a commutative diagram of homomorphisms

$$
\begin{array}{ccc}
R & \longrightarrow & S \\
 & \searrow & \downarrow \\
 & & T
\end{array}
$$

and isomorphisms $A \otimes_R S \cong B$ and $A \otimes_R T \cong C$, there exists a unique morphism $B \to C$ completing the diagram:

$$
B \otimes_S T \cong A \otimes_R S \otimes_S T = A \otimes_R T \cong C \, .
$$

A category with these properties goes under the name of cofibred groupoid (the dual functor on germs of spaces is a fibred groupoid). We give the definition, following [Bi1]; it is equivalent to the definition in [Rim1].

Definition. A *cofibred groupoid* over a category C is a category F equipped with a covariant functor $p: F \to C$, satisfying

1. Existence of liftings: for every morphism $\Phi: R \to S$ in C and every object r in F with $p(r) = R$ a lift $\phi: r \to s$ exists with $p_*(\phi) = \Phi$,
2. Uniqueness: for every commutative diagram

$$R \longrightarrow S$$
$$\searrow \quad \downarrow$$
$$T$$

and for every lift $r \to s$ and $r \to t$ there is a unique $s \to t$ over $S \to T$ making the diagram commutative:

$$r \longrightarrow s$$
$$\searrow \quad \downarrow!$$
$$t$$

Definition. For an object R in C the fibre $F(R)$ is the subcategory of F, whose objects lie over R, while the morphisms lie over id_R.
If s is an object of F, then F_s is the category, whose objects are the morphisms $r \to s$ in F, and morphisms are commutative diagrams

$$r \longrightarrow r'$$
$$\searrow \quad \downarrow$$
$$s$$

Remark. The fibre $F(R)$ is a groupoid in the usual sense, a category in which all morphisms are isomorphisms.

Remark. For a fixed $\mathbb{C}$-algebra A_0 the deformation groupoid $\mathrm{Def}_{A_0/\mathbb{C}}$ has, according to the definition above, as objects triples (A, R, i_A), consisting of a flat R-algebra A, and an isomorphism $i_A \colon A/\mathfrak{m}_R A \to A_0$. A morphism $(A, R, i_A) \to (B, S, i_B)$ over $R \to S$ is given by an S-algebra isomorphism $\phi \colon A \otimes_R S \to B$, such that the following diagram commutes:

$$A \otimes_R \mathbb{C} \xrightarrow{\phi \otimes \mathbb{C}} B \otimes_S \mathbb{C}$$
$$\searrow \qquad \swarrow$$
$$A_0$$

But this is exactly what we defined to be a deformation of A_0.

Let $\overline{F}(R)$ be the set of isomorphism classes in $F(R)$. Then

$$\overline{F} \colon C \to (\mathrm{Sets})^{\dagger}, \qquad R \mapsto \overline{F}(R)$$

is a functor in a natural way. Similarly, $\overline{F}_a(R)$ is the set of isomorphism classes in $F_a(R)$. In particular, $\overline{\mathrm{Def}}_{A_0/\mathbb{C}}$ is the deformation functor, as it is

---

† This is not the category of all sets, but of sets in a certain universe, see [SGA I, p. 146].

often introduced in the literature. For most purposes this definition suffices. Because of its importance, we give it here explicitly.

Let A_0 be an analytic algebra, or consider dually a singularity $(X_0, 0)$. In the last case it is more natural to look at the functor on the category (GAn) of germs of analytic spaces. The functor of deformations of X_0 is (with a slightly simplified notation) the functor

$$\overline{\mathrm{Def}}_{X_0} \colon (\mathrm{GAn}) \to (\mathrm{Sets})$$

$$\overline{\mathrm{Def}}_{X_0}(T) = \{ \text{Isomorphism classes of deformations of } X_0 \text{ over } T \}.$$

By abuse of notation $\overline{\mathrm{Def}}_{X_0}$ is also used for the functor

$$\overline{\mathrm{Def}}_{X_0} \colon (\mathrm{Analg}) \to (\mathrm{Sets})$$

$$\overline{\mathrm{Def}}_{X_0}(R) = \{ \text{Isomorphism classes of deformations of } A_0 \text{ over } R \}.$$

Returning to the deformation groupoid, we remark that the fibre $\overline{\mathrm{Def}}_{A_0/\mathbb{C}}(\mathbb{C})$ contains only one isomorphism class; more generally, we will always assume that $\overline{F}_r(\mathbb{C})$ is a category in which $\mathrm{Hom}_{F_r(\mathbb{C})}(a, b)$ consists of exactly one element for any two objects a and b.

Definition. Let F be a cofibred groupoid over the category (Analg). An object $s \in F(S)$ is *versal*, if it satisfies the following lifting property:

> Let $r' \to r$ be a morphism in F with $R' \to R$ surjective. Then every morphism $s \to r$ can be lifted to a morphism $s \to r'$, such that $(s \to r' \to r) = (s \to r)$.

The object s is *formally versal*, if the lifting property is satisfied for Artinian algebras R'. The object S is *(formally) semi-universal*, if it is (formally) versal, and if the induced morphism $m_S/m_S^2 \to m_{R'}/m_{R'}^2$ is uniquely determined by $r' \to r$ and $s \to r$. We also use the term *miniversal* for semi-universal. Finally S is *universal* if the map $S \to R'$ itself is unique.

Remark. The treatment in [Schl1] and other sources is slightly more general, in the sense that the base category is that of algebras over a fixed base ring Λ; the condition for miniversality then involves $m_S/(m_\Lambda S + m_S^2)$. This is the most important change, otherwise one has just the extra Λ in the notation.

Furthermore, these sources deal with the algebraic case: let Λ be a complete Noetherian local ring with residue field k and let $\mathcal{C}_\Lambda$ be the category of Artinian local Λ-algebras with the same residue field k. The category $\mathcal{C}_\Lambda$ is a full subcategory of the category $\widehat{\mathcal{C}}_\Lambda$ of complete Noetherian local Λ-algebras with the residue field k. A cofibred groupoid F over $\mathcal{C}_\Lambda$ extends to a groupoid $\widehat{F}$ over $\widehat{\mathcal{C}}_\Lambda$, by putting $\widehat{F}(R) = \varprojlim F(R/m_r^k)$, and for a groupoid F over $\widehat{\mathcal{C}}_\Lambda$ one obtains a functor $\widehat{} \colon F \to \widehat{F}$. Objects of $\widehat{F}$ are called *formal* objects. The definitions of versal and formally versal extend in an obvious way to these groupoids. Deformations are again defined as flat morphisms.

Remark. In the literature one often finds a definition of versality, which requires only the lifting property for $r = s \otimes \mathbb{C}$, or in other words: every deformation is induced from a versal one. Such a definition is not strong enough to prove that a versal deformation is the product of a miniversal one with a trivial factor. As example of a 'versal' deformation in this weak sense, which is not versal, one can take the union of versal deformation with any other deformation. We can then induce from the versal part.

The construction of versal objects can be divided in three steps:

1. Existence of a formally versal, formal object $\varprojlim s_m$ over a complete local ring $\overline{S} = \varprojlim S_m$. Criteria for this are given by the Schlessinger conditions, discussed below.
2. The formal object $\varprojlim s_m$ in $\widehat{F}(\overline{S})$ is not necessarily an object in $F(\overline{S})$. One shows the existence of an $\overline{s} \in F(\overline{S})$, which induces $\varprojlim s_m$. By some variant of Artin approximation one finds an analytic algebra S and an analytic object $s \in F(S)$, which is formally versal. These are really two steps, formulated in one, because sometimes s and S can directly be found from the result of the first part, as in GRAUERT's proof of the existence of versal deformations for isolated singularities (cf. Chap. 9).
3. The analytic, formally versal object $s \in F(S)$ is versal. For conditions on F, under which formal versality already implies versality, see [Bi1, Sect. 3]. These conditions are satisfied for most known deformation problems, in particular for deformations of compact complex spaces, of isolated singularities, of coherent modules, and of principal bundles on compact complex spaces.

Example. In the following example step 3 cannot be done; it is a slight modification of an example due to BUCHWEITZ. Consider deformations of a smooth affine curve; to be specific, take $C: x^3 + y^3 + 1 = 0$. Every deformation over a zero-dimensional base is trivial ($T_C^1 = 0$ because C is smooth), so $C \to 0$ is an analytic, even algebraic, formally versal object. Nevertheless the one-parameter family

$$x^3 + y^3 + 1 + \lambda xy$$

is non-trivial (it is the Hesse normal form for cubic curves in affine coordinates, with λ the modular parameter).

We first show that the family is trivial modulo (λ^2). Consider the element $xy \in H^0(C, N) = \operatorname{Im} H^0(\Theta_{\mathbb{C}^2}|_C)$. We determine a, b, and c, such that

$$3ax^2 + 3by^2 + c(1 + x^3 + y^3) = -xy .$$

We can take $c = -xy$, $a = 1/3x^2y$, and $b = 1/3xy^2$. Then

$$(1 - \lambda xy)\left((x + \tfrac{1}{3}\lambda x^2 y)^3 + (y + \tfrac{1}{3}\lambda xy^2)^3 + 1 + \lambda(x + \tfrac{1}{3}\lambda x^2 y)(y + \tfrac{1}{3}\lambda xy^2)\right)$$
$$\equiv 1 + x^3 + y^3 \pmod{\lambda^2} .$$

This computation also shows what to do in general: put $t = \lambda xy$, and find (formal) power series $\varphi(t)$ and $\psi(t)$, with $\varphi(0) = \psi(0) = 1$, such that the coordinate transformation $(x,y) \mapsto (x\varphi, y\varphi)$ trivialises the family:

$$\psi(x^3\varphi^3 + y^3\varphi^3 + 1 + t\varphi^2) = x^3 + y^3 + 1 \ .$$

This leads to the two equations $\psi\varphi^3 = 1$, $\psi(1 + t\varphi^2) = 1$, from which we get

$$\left(\varphi(t)\right)^3 - t\left(\varphi(t)\right)^2 - 1 = 0 \ .$$

The function φ is the inverse of the rational function $t = z - 1/z^2$ in the neighbourhood of $(z,t) = (1,0)$, and is as such an analytic function with finite radius of convergence K. Therefore the coordinate transformations above converge for $|\lambda xy| < K$, so there is no value of the parameter λ with convergence on the whole affine plane.

Example: the algebraic case. One cannot expect the existence of an affine versal deformation of affine schemes with isolated singularities. Instead one has to work with étale neighbourhoods of the singular locus; the base ring of a versal deformation is the henselisation of a k-algebra of finite type. By a result of ELKIK [El], algebraic versal deformations of isolated singularities in this sense exist. For more details, see [Art3, KPR].

For deformations of proper schemes it can happen that an algebraic versal deformation does not exist: an algebraic $K3$ surface over $\mathbb{C}$ has, considered as compact complex manifold, an (analytic) versal deformation with smooth base space of dimension 20, but algebraic $K3$ surfaces form only 19 dimensional families.

We now come to SCHLESSINGER's conditions. Let F be a cofibred groupoid over the category (Analg) and denote also by F the cofibred groupoid over the full subcategory $\mathcal{C}$ of Artinian algebras.

Definition. A surjection $\Phi : R' \to R$ in $\mathcal{C}$ is a *small extension*, if $(\ker \Phi)^2 = 0$ and $\ker \Phi \cong \mathbb{C}$, i.e., if for some $t \in R'$ with $t^2 = 0$ the sequence

$$0 \longrightarrow \mathbb{C} \xrightarrow{\cdot t} R' \xrightarrow{\Phi} R \longrightarrow 0$$

is exact.

Condition (S1). Let $R' \to R$ and $R'' \to R$ be morphisms in $\mathcal{C}$, let $r \in F(R)$ and consider the canonical map $\overline{F}_r(R' \times_R R'') \to \overline{F}_r(R') \times_{\overline{F}_r(R)} \overline{F}_r(R'')$:

(a) this map is a surjection if $R'' \to R$ is a small extension.
(b) it is bijective if $R = \mathbb{C}$ and $R'' = \mathbb{C}[\varepsilon]$.

This condition depends only on the functor $\overline{F}$; it is implied by the following stronger condition on F, which is satisfied in many applications:

Condition (S1'). In the situation of condition (S1)(a) the map

$$F_r(R' \times_R R'') \to F_r(R') \times_{F_r(R)} F_r(R'')$$

is an equivalence of categories.

This condition involves the fibre product of categories, which has to be interpreted in the correct sense (see [Rim1, 2.6]), so we give the less nice formulation of what it actually means [Rim1, 2.5]:

Condition (S1'). Let $R'' \to R$ be a small surjection, and consider $r \in F(R)$. For $(r' \to r) \in F_r(R')$ and $(r'' \to r) \in F_r(R'')$ the fibre product $(r' \times_r r'' \to r) \in F_r(R' \times_R R'')$ exists.

Remark. A functor satisfying condition (S1) is also called *semi-homogeneous*, while the name *homogeneous* refers to condition (S1'). The conditions (S1)(a) and (S1)(b) are called (H1) and (H2) in [Schl1], from *hull*, which is SCHLESSINGER's terminology for a formally miniversal formal object. I follow here (qua notation) [Art1], which introduced generalised Schlessinger conditions, for cofibred groupoids over the category of Noetherian Λ-algebras; cf. also [Bi1, Fl] for the dual formulation of fibred groupoids over the category (An) of analytic spaces.

Remark. By Condition (S1)(b) one gets a natural vector space structure on $\overline{F}_r(\mathbb{C}[\varepsilon])$ for $r \in F(\mathbb{C})$; the multiplication with $c \in \mathbb{C}^*$ comes from the isomorphism $\mathbb{C}[\varepsilon] \to \mathbb{C}[\varepsilon] \colon \varepsilon \mapsto c\varepsilon$, which induces an bijection $c_* \colon F(\mathbb{C}[\varepsilon]) \to F(\mathbb{C}[\varepsilon])$, and the addition from the map $\mathbb{C}[\varepsilon] \times_{\mathbb{C}} \mathbb{C}[\varepsilon]$, given by $(a+b\varepsilon, a+c\varepsilon) \mapsto (a, (b+c)\varepsilon)$. Therefore the following condition makes sense:

Condition (S2). Let $r \in F(\mathbb{C})$. Then $\dim_{\mathbb{C}} \overline{F}_r(\mathbb{C}[\varepsilon]) < \infty$.

Let F be a cofibred groupoid over the category $\mathcal{C}$, which satisfies the conditions (S1) and (S2). For every $r \in F(\mathbb{C})$ a formal miniversal object in $\widehat{F}$ exists. This follows from:

Theorem (Schlessinger). *Let $\overline{F} \colon \mathcal{C} \to$ (Sets) be a functor, for which $\overline{F}(\mathbb{C})$ consists of one element. A formally miniversal object exists if and only if $\overline{F}$ satisfies (S1) and (S2).*

Proof. The necessity is easy to see, we only give the construction. Let $\dim_{\mathbb{C}} \overline{F}(\mathbb{C}[\varepsilon]) = k$, and take a basis $(\vartheta_1, \dots, \vartheta_k)$. We construct S as limit of successive quotients of the formal power series ring $P = \mathbb{C}[[T_1, \dots, T_k]]$. We determine inductively $S_m = P/J_m$, and an $s_m \in \overline{F}(S_m)$, which satisfies the lifting property for all Artinian rings R' with $\mathfrak{m}_{R'}^{m+1} = 0$.

Put $S_1 = P/\mathfrak{m}^2 \cong \mathbb{C}[\varepsilon] \times \cdots \times \mathbb{C}[\varepsilon]$, and let $s_1 \in \overline{F}(S_1)$ be the element, corresponding to $\vartheta_1 \times \cdots \times \vartheta_k$.

Let $s_m \in \overline{F}(S_m)$ be constructed. The set of ideals J with the property that $\mathfrak{m}_P J_m \subset J \subset J_m$, and that s_m can be lifted to an object in $\overline{F}(P/J)$, is closed under intersection: if J and K are such ideals, we can, by enlarging J without changing $J \cap K$, assume that $J + K = J_m$; as $P/J \times_{P/J_m} P/K \cong P/(J \cap K)$, Condition (S1)(a) gives a lift to $P/(J \cap K)$. Let J_{m+1} be the minimal element, put $S_{m+1} = P/J_{m+1}$ and take for $s_{m+1} \in \overline{F}(S_{m+1})$ any lift of s_m.

We have to show that the formal object $\varprojlim s_m$ satisfies the lifting property for Artinian algebras; by induction on the length it suffices to consider a small surjection $R' \to R$; let $t \in R'$ be an element with $\mathfrak{m}_{R'} t = 0$. Let $r' \to r$ lie over $R' \to R$. It suffices to complete the top line of every diagram

$$
\begin{array}{ccc}
s_{m+1} & \dashrightarrow & r' \\
\downarrow & & \downarrow \\
s_m & \longrightarrow & r
\end{array}
$$

Define $S'_m := S_m \times_R R'$ and set $u' = (0, t)$, then $S'_m/u'S'_m = S_m$ and $\mathfrak{m}_{S'_m} u' = 0$. Condition (S1)(a) gives an object $s'_m \in \overline{F}(S'_m)$, mapping to r' and s_m.

As P is a power series ring, the map $P \to S_m$ can be lifted to a map $P \to S'_m$. Consider the diagram

$$
\begin{array}{ccccc}
P & \xrightarrow{w} & S'_m & \xrightarrow{p} & R' \\
\downarrow & & \downarrow q & & \downarrow \\
S_{m+1} & \longrightarrow & S_m & \longrightarrow & R
\end{array}
$$

There are now two possibilities: either the map w is surjective, in which case $J_{m+1} \subset \ker w$ by definition of J_{m+1} (because a lift s'_m exists), and w factorises over S_{m+1}, or the projection q maps $w(P)$ isomorphically onto S_m (for dimension reasons), and this map can be inverted. In both cases there is a map $v \colon S_{m+1} \to S'_m$ such that r' and $r'' := (p \circ v)_* s_{m+1}$ have the same image r. But then we are done, because $\overline{F}(\mathbb{C}[\varepsilon])$ acts transitively on the fibre over r. For this we remark that $R' \times_R R'$ is isomorphic to $R' \times \mathbb{C}[\varepsilon]$ under the map $(x, y) \mapsto (x, \overline{x} + (y - x))$, and by (S1)(b) and (S1)(a) we have a surjection

$$
\overline{F}(R') \times \overline{F}(\mathbb{C}[\varepsilon]) \cong \overline{F}(R' \times \mathbb{C}[\varepsilon]) \cong \overline{F}(R' \times_R R') \longrightarrow \overline{F}(R') \times_{\overline{F}(R)} \overline{F}(R') .
$$

□

A stronger property is that a universal family exists. Then there is a base-ring S and the functor $\overline{F}$ satisfies $\overline{F}(A) = \mathrm{Hom}(S, A) =: h_S(A)$. When this last property holds for Artinian A the functor $\overline{F}$ is called *pro-representable*. A sufficient condition is that (S1) and (S2), or (H1)–(H3), hold together with the fourth Schlessinger condition

Condition (H4). For a small extension $R' \to R$ one has $\overline{F}(R' \times_R R') \cong \overline{F}(R') \times_{\overline{F}(R)} \overline{F}(R')$.

Theorem. *The deformation groupoid* Def *is homogeneous. For isolated singularities formal versal deformations exist.*

Proof. We check condition (S1′). As we use the letters F and R to describe functions and relations, we consider rings $A' \xrightarrow{\Phi} A \longleftarrow A''$, with $\rho\colon A'' \to A$ a small extension. We write the elements of A'' as $a + bt$, so $\rho(a + bt) = a$. Then

$$A' \times_A A'' = \{\, (a', \Phi(a') + bt) \mid a' \in A, b \in \mathbb{C} \,\} \,.$$

A singularity X'' over A'' is given by equations of the form $F(x, a) + tF''(x)$, and relations $R(x, a) + tR''(x)$, such that

$$
\begin{aligned}
0 = {} & \rho(F(x, a)R(x, a)) \\
& + t\big(F(x, a)R(x, a) - \rho(F(x, a)R(x, a)) + F(x, 0)R''(x) + F''(x)R(x, 0)\big)
\end{aligned}
$$

in $A''[[x]]$. Consider a deformation X' over A', mapping to $\mathcal{O}_{X''} \otimes A$ under a map induced from $\phi\colon A'[[x]] \to A[[x]]$. Using an automorphism of $A[[x]]$ we may assume that ϕ is given by applying Φ to the coefficients. The images of the generators of the ideal of X' are expressible in the equations $F(x, a)$, and coincide with them when reduced to $\mathbb{C} = A/\mathfrak{m}_A = A'/\mathfrak{m}_{A'}$. So we may assume that the equations $F'(x, a')$ satisfy $\phi(F'(x, a')) = F(x, \Phi(a'))$, and likewise for the relations $\phi(R'(x, a')) = R(x, \Phi(a'))$.

We now define a deformation over $A' \times_A A''$ by the equations

$$\big(F'(x, a'), F(x, \Phi(a')) + tF''(x)\big) \,.$$

A relation matrix is then $\big(R'(x, a'), R(x, \Phi(a')) + tF''(x)\big)$.

The vector space $\overline{\mathrm{Def}}_{X_0}(\mathbb{C}[\varepsilon])$ is the space $T^1_{X_0}$ of first order infinitesimal deformations, and we have seen that this is finite dimensional for isolated singularities. □

Openness of versality. Consider a flat holomorphic map $f\colon X \to S$, such that the critical space Σ of f is finite over S. Openness of versality is the property that the set of points $s \in S$, such that $(X, \Sigma_s) \to (S, s)$ is a versal deformation of the germ $(X_s, \Sigma_s) \to (S, s)$, is Zariski open. For a germ $f\colon (X, 0) \to (S, 0)$ this applies for a suitable representative. The existence proof of versal deformations with DOUADY's Banach-analytic methods [Po] also shows this property; this is also true for other deformation problems, in contrast with the power series method. ARTIN has quite generally shown that (formal) versality is an open property for 'sufficiently nice' deformation problems in algebraic geometry [Art1]. BINGENER has extended these results to the analytic case [Bi1]; for a somewhat simpler proof see [Fl].

In order to formulate the statements, one has to consider more global objects; this leads to cofibred categories over the category (An) of analytic spaces, or a category of schemes, alluded to above. Then one defines some more conditions, which I will not give here; they involve a functorial *obstruction theory*. For deformations of compact complex manifolds and of isolated singularities the obstruction theory is provided by the *cotangent complex* (see [Bi1, Pa1, Pa3]). For these deformation problems formal versality is an open property. To conclude that also versality is open, one can use the following useful criterion of FLENNER:

Proposition [Fl, Satz (5.2)]. *Suppose a versal deformation $s \in F$ of $s_0 \in F(\mathbb{C})$ exists. Then also a miniversal deformation of s_0 exists in F, and every formally versal deformation of s_0 in F is versal.*

The proof uses the following lemma, whose proof we include, because it is instructive.

Lemma [Fl, Lemma (5.3)]. *Let $\bar{r} \to \bar{s}$ be a morphism in $\widehat{F}$, with $\bar{r}$ formally miniversal and $\bar{s}$ formally versal. Then is $\overline{S}$ a free power series ring over $\overline{R}$.*

Proof. By versality of $\bar{s}$ there is a map $\bar{s} \to \bar{r}$, and the composition $\bar{s} \to \bar{r} \to \bar{s}$ induces the identity map $\mathfrak{m}_{\overline{R}}/\mathfrak{m}_{\overline{R}}^2 \to \mathfrak{m}_{\overline{R}}/\mathfrak{m}_{\overline{R}}^2$, by formal miniversality. In particular, $\mathfrak{m}_{\overline{R}}/\mathfrak{m}_{\overline{R}}^2 \to \mathfrak{m}_{\overline{S}}/\mathfrak{m}_{\overline{S}}^2$ is injective. The difference in dimension equals $n = \dim_{\mathbb{C}}(\mathfrak{m}_{\overline{S}}/\mathfrak{m}_{\overline{R}}\overline{S} + \mathfrak{m}_{\overline{S}}^2)$. There is a surjective $\overline{R}$-homomorphism $\psi: \overline{Q} := \overline{R}[[X]] \to \overline{S}$, inducing an isomorphism $\overline{Q}_1 := \overline{Q}/\mathfrak{m}_{\overline{Q}}^2 \xrightarrow{\sim} \overline{S}/\mathfrak{m}_{\overline{S}}^2 =: \overline{S}_1$. As $\bar{s}$ is formally versal, the top line of the diagram

$$
\begin{array}{ccc}
\bar{s} & \dashrightarrow & \bar{r} \otimes_{\overline{R}} \overline{C} \\
\downarrow & & \downarrow \\
\bar{s} \otimes_{\overline{S}} \overline{S}_1 & \xrightarrow{\sim} & \bar{r} \otimes_{\overline{R}} \overline{C}_1
\end{array}
$$

can be completed, and induces a $\mathbb{C}$-homomorphism $\varphi: \overline{S} \to \overline{Q}$, which is surjective by construction. The compositions $\varphi\psi$ and $\psi\varphi$ are as surjective endomorphisms already isomorphisms, so ψ is an isomorphism. $\qquad\square$

As corollary one obtains that the base ring of a versal deformation is a free convergent power series ring over the base space of a miniversal deformation (if this exists).

7 Deformations of compact manifolds

The first modern deformation theory was the Kodaira–Spencer theory of deformations of compact complex manifolds; for a vivid account of its development, see KODAIRA's book [Kod].

A deformation $\pi: \mathcal{M} \to S$ of a smooth compact manifold M over a pointed base space $(S, 0)$ is by definition a proper flat map, with M isomorphic to the fibre $\pi^{-1}(0)$. If S is smooth, the flatness condition means that π is a submersion.

Although in concrete cases it is often not the most useful description, we view a complex manifold M as obtained by gluing domains in $\mathbb{C}^n$. Let $\mathfrak{U} = \{\mathcal{U}_i \mid i = 1, 2, \ldots, l\}$ be a finite cover of M, with each $\mathcal{U}_i$ isomorphic to a unit polydisc U_i via a local coordinate $z_i = (z_i^1, \ldots, z_i^n)$; we consider z_i either as map $z_i: \mathcal{U}_i \to U_i$, or as a point of U_i. If $p \in M$ belongs to the open set $\mathcal{U}_i \cap \mathcal{U}_j$, we can use two coordinate functions, so $z_i(p) = f_{ij}(z_j(p))$, where f_{ij} is a holomorphic function on $U_{ij} := z_j(\mathcal{U}_i \cap \mathcal{U}_j) \subset U_j$.

The fundamental idea of KODAIRA and SPENCER is to consider a deformation of M to be the gluing of the same polydiscs U_i, but with different identification, cf. [Kod, 4.1.(a)]. Consider a deformation $\pi: \mathcal{M} \to S$ with smooth base. If S is sufficiently small, we have $\mathcal{M} = \cup_{i=1}^l U_i \times S$. The points $(z_i, t) \in U_j \times S$ and $(z_j, t) \in U_j \times S$ represent the same point of $\mathcal{M}$, if $z_i = f_{ij}(z_j, t)$. It is important to remark that the domain of the holomorphic function f_{ij} depends on t. If $\mathcal{U}_i \cap \mathcal{U}_j \cap \mathcal{U}_k \neq \emptyset$, the following relation holds:

$$f_{ik}(z_k, t) = f_{ij}(f_{jk}(z_k, t), t) .$$

To find the induced infinitesimal deformation, we derive this equation with respect to t; for this we first suppose that S is one-dimensional. We get

$$\frac{\partial f_{ik}}{\partial t} = \frac{\partial f_{ij}}{\partial t} + \frac{\partial z_i}{\partial z_j} \frac{\partial f_{jk}}{\partial t} , \tag{7.1}$$

where $\partial z_i / \partial z_j$ is the Jacobi matrix. We consider this as equation between tangent vectors. Using the explicit basis $\partial / \partial z_i^\alpha$ of the tangent space we introduce the vector field

$$\theta_{ij}(t) = \frac{\partial f_{ij}}{\partial t} \cdot \frac{\partial}{\partial z_i} = \sum \frac{\partial f_{ij}^\alpha(z_j, t)}{\partial t} \frac{\partial}{\partial z_i^\alpha} ,$$

which is defined on $\mathcal{U}_i \cap \mathcal{U}_j$. The equation (7.1) now reads

$$\theta_{jk}(t) - \theta_{ik}(t) + \theta_{ij}(t) = 0 \; , \tag{7.2}$$

so the $\theta_{ij}(t)$ define a cocycle on $\mathfrak{U}_t$, and its cohomology class $\theta(t)$ in $H^1(\mathfrak{U}_t, \Theta_t) = H^1(\mathcal{M}_t, \Theta_t)$ is the *Kodaira-Spencer class* of the deformation. One checks that this cohomology class does not depend on the choice of systems of local coordinates; by taking refinements of the covering one reduces to the case of different coordinates (w_i, t) on the same $\mathcal{U}_i$. Let $w_i = h_{ij}(w_j, t)$ and $w_i = g_i(z_i, t)$. Then on $\mathcal{U}_i \cap \mathcal{U}_j$

$$g_i(f_{ij}(z_j, t), t) = h_{ij}(g_j(z_j, t), t) \; ,$$

and therefore

$$\frac{\partial g_i}{\partial z_i} \frac{\partial f_{ij}}{\partial t} + \frac{\partial g_i}{\partial t} = \frac{\partial h_{ij}}{\partial w_j} \frac{\partial g_j}{\partial t} + \frac{\partial h_{ij}}{\partial t} \; .$$

With the vector fields $\eta_{ij} = \partial h_{ij}/\partial t \cdot \partial/\partial w_i$ and $\theta_i = \partial g_i/\partial t \cdot \partial/\partial w_i$ we rewrite this equation as

$$\theta_{ij} - \eta_{ij} = \theta_j - \theta_i \; ,$$

so θ_{ij} and η_{ij} differ by a coboundary.

For a family $\pi \colon \mathcal{M} \to S$ over a higher dimensional base S we obtain for each $t \in S$ the *Kodaira–Spencer map* $\kappa_t \colon T_t S \to H^1(\mathcal{M}_t, \Theta_t)$ by the formula

$$\kappa_t \left(\sum c_\alpha \frac{\partial}{\partial t_\alpha} \right) = \sum c_\alpha \frac{\partial f_{ij}}{\partial t_\alpha} \cdot \frac{\partial}{\partial z_i} \; .$$

Conversely, given a class $\theta_{ij} \in H^1(M, \Theta)$ the existence problem asks for a one-parameter deformation $\pi \colon \mathcal{M} \to S$ of M with θ_{ij} as Kodaira–Spencer class. Obstructions can appear. To find them, we suppose that a deformation exists, and derive (7.2) again with respect to t. Let $\theta_{ij} = \sum_\alpha \theta_{ij}^\alpha(z_i, t)\partial/\partial z_i^\alpha$, and define the new vector field

$$\dot{\theta}_{ij}(t) = \sum_\alpha \frac{\partial \theta_{ij}^\alpha(z_i, t)}{\partial t} \frac{\partial}{\partial z_i^\alpha} \; .$$

This definition involves local coordinates, so we have to do the calculation in one system of coordinates, for which we take z_j. Because the vector field θ_{ij} is the derivative of the transformation from z_j to z_i, we obtain after a small computation

$$\frac{\partial}{\partial t}\theta_{ik}(t) = \dot{\theta}_{ik}(t) + [\theta_{ij}(t), \theta_{ik}(t)] \; ,$$

and deriving (7.2) we find

$$\dot{\theta}_{jk}(t) - \dot{\theta}_{ik}(t) + \dot{\theta}_{ij}(t) = [\theta_{ij}(t), \theta_{jk}(t)] \; .$$

We reformulate this equation with the bracket on the Čech cohomology $H^*(M,\Theta)$, which is a graded Lie-algebra under the operation $H^p(M,\Theta) \otimes H^q(M,\Theta) \to H^{p+q}(M,\Theta)$, defined by

$$[\vartheta, \psi]_{0\ldots p+q} = [\vartheta_{0\ldots p}, \psi_{p\ldots p+q}] .$$

With this operation our equation becomes

$$\delta\dot{\theta}(t) = [\theta(t), \theta(t)] ,$$

and we have shown:

Proposition. *A necessary condition for the existence of a one parameter deformation with Kodaira-Spencer class $\theta \in H^1(M,\Theta)$ is that $[\theta, \theta] = 0$ in $H^2(M,\Theta)$.*

One could continue, and find higher obstructions, which also lie in $H^2(M,\Theta)$. Even if $H^2(M,\Theta) = 0$, and all obstructions vanish, it is not easy to find a convergent solution to the problem. For this one needs to shrink the polydiscs: although one can define formal transition functions $F_{ij} = \sum F_{ij}^\alpha t^\alpha$ on U_{ij}, a convergent solution will not map U_{ij} to U_{ji}. All one can say is that for small t the image will lie in a small neighbourhood of U_{ji}, or that a slightly smaller subset of U_{ij} maps into U_{ji}. For a proof of the existence of a miniversal deformation, based on this direct approach, see [Com].

KODAIRA and SPENCER have used a different method for the existence theorem. Along these lines KURANISHI later proved the existence of versal deformations (cf. [Ku]). In a family $\mathcal{M} \to S$ over a reduced base all fibres $\mathcal{M}_t$ are diffeomorphic to $M = \mathcal{M}_0$: we can view the deformation also as a family of complex structures on the same underlying differentiable manifold M. An almost complex structure on M is an automorphism J of the tangent bundle TM with $J^2 = -\mathrm{Id}$. The complexified tangent bundle $\mathbb{C}TM$ splits as direct sum of the $+i$ and $-i$ eigenspaces $T' \cong \Theta$ (the holomorphic tangent bundle) and T''. The latter is of the form $\{X + iJ(X) \mid X \in TM\}$. Conversely, knowledge of T'' allows to find J. Now suppose that the almost complex structure $J(t)$ is a deformation of a given complex structure J_0. For small t the space $T''(t)$ can be described as graph of a map $T'' \to T'$, where T' and T'' come from our fixed structure J_0, or in other words as a T'-valued (and therefore Θ-valued) differential form $\vartheta(t)$ of type $(0,1)$. The structure is integrable if and only if

$$\bar{\partial}\vartheta(t) + \tfrac{1}{2}[\vartheta(t), \vartheta(t)] = 0 ,$$

where this time the bracket is the bracket on the Dolbeault complex, which computes $H^1(M,\Theta)$. This deformation equation has now to be solved (cf. the next Chapter).

Example: Deformations of projective hypersurfaces. This basic example can already be found in the classic paper [KS]. A smooth hypersurface $M \subset \mathbb{P}^{n+1}$

of degree d can be described by a homogeneous polynomial $F(x_0, \ldots, x_{n+1})$ of degree d, whose partial derivatives have no common zeroes on $\mathbb{P}^{n+1}$. Let $\mathbb{P} := \mathbb{P}\left(H^0\left(\mathbb{P}^{n+1}, \mathcal{O}(d)\right)\right)$ be the space of all hypersurfaces. The second projection exhibits the incidence variety $\mathcal{V} = \{(x,p) \in \mathbb{P}^{n+1} \times \mathbb{P} \mid p(x) = 0\}$ as total space of the universal family $p_2 \colon \mathcal{V} \to \mathbb{P}$ of hypersurfaces. Let $\Delta \subset \mathbb{P}$ be the discriminant, and set $S = \mathbb{P} \setminus \Delta$. Define $\mathcal{M} = p_2^{-1}(S)$.

Proposition. *The family $\mathcal{M} \to S$ of smooth hypersurfaces of dimension $n > 1$ is a versal deformation of each of its fibres with the exception of the case $n = 2$, $d = 4$.*

Proof. Let M be one of the fibres. Consider the normal bundle sequence of M:

$$0 \longrightarrow \Theta_M \longrightarrow \Theta_{\mathbb{P}^{n+1}}|_M \longrightarrow \mathcal{O}_M(d) \longrightarrow 0 \,.$$

Generators of the $\mathbb{C}$-vector space $H^0(\mathcal{O}_M(d))$ are the monomials of degree d, and there is one relation, coming from the defining equation of M; $H^0(\mathcal{O}_M(d))$ is the tangent space to S. The deformation is versal, if the coboundary map to $H^1(\Theta_M)$ is surjective; to prove this statement, one has to give an explicit description of this map, and compare it with the definition of the Kodaira-Spencer class, which is done in detail in [Kod, Sect. 5.2]. We compute $H^1(\Theta_{\mathbb{P}^{n+1}}|_M)$ with standard methods. Consider the exact sequence

$$0 \longrightarrow \Theta_{\mathbb{P}^{n+1}}(-d) \longrightarrow \Theta_{\mathbb{P}^{n+1}} \longrightarrow \Theta_{\mathbb{P}^{n+1}}|_M \longrightarrow 0 \,,$$

and, in order to find $H^i(\Theta_{\mathbb{P}^{n+1}}(-m))$, $m \geq 0$, $i > 0$, also the twisted Euler sequence

$$0 \longrightarrow \mathcal{O}_{\mathbb{P}^{n+1}}(-m) \longrightarrow \bigoplus_{n+2} \mathcal{O}_{\mathbb{P}^{n+1}}(1-m) \longrightarrow \Theta_{\mathbb{P}^{n+1}}(-m) \longrightarrow 0 \,.$$

We write shortly Θ for $\Theta_{\mathbb{P}^{n+1}}$ and $\mathcal{O}$ for $\mathcal{O}_{\mathbb{P}^{n+1}}$. Because $H^i(\mathcal{O}(r)) = 0$ for $0 < i < n+1$ [Ha, III.5.1], we are reduced to considering the following sequence:

$$0 \longrightarrow H^n(\Theta(-m)) \longrightarrow H^{n+1}(\mathcal{O}(-m)) \longrightarrow \bigoplus_{n+2} H^{n+1}(\mathcal{O}(1-m))$$
$$\longrightarrow H^{n+1}(\Theta(-m)) \longrightarrow 0 \,.$$

By Serre duality $H^n(\Theta(-m))$ is dual to the cokernel of the map

$$\bigoplus_{n+2} H^0(\mathcal{O}(m-n-3)) \to H^0(\mathcal{O}(m-n-2)) \,,$$

which comes from the multiplication map $H^0(\mathcal{O}(1)) \otimes \mathcal{O}(m-n-3) \to \mathcal{O}(m-n-2)$; it is surjective, except when $m = n+2$. Combining everything we find $H^1(\Theta_{\mathbb{P}^{n+1}}|_M) = 0$, except for $n = 2$, $d = 4$. $\square$

In the survey [Pa3] PALAMODOV extends this result to complete intersections: the universal family is versal for its fibres, except for surfaces, which are the intersection of a quadric and a cubic, or the intersection of three quadrics. In the exceptional cases the surfaces in question are $K3$-surfaces. The moduli space of $K3$-surfaces is 20 dimensional, but the 'general' surface is not algebraic. Only if one fixes a polarisation, which is more or less equivalent to fixing a projective embedding, one obtains algebraic families, of dimension 19. The deformation functor satisfies Schlessinger's conditions, so a formal versal deformation exists, which is even analytic, but it cannot be made algebraic.

Example: Hopf surfaces. This example was important in the development of the theory [KS], because although the base space of the versal deformation is still smooth, the dimension of $H^1(\Theta_{M_t})$ is not constant. This means that a moduli space of Hopf surfaces does not exist in any reasonable sense: the space of isomorphism classes cannot be Hausdorff.

Let A be an $n \times n$ matrix, which operates as contraction on $\mathbb{C}^n$, so the eigenvalues satisfy $0 < |\lambda| < 1$. The action of the infinite cyclic group G_A, generated by A, on $\mathbb{C}^n \setminus \{0\}$ is fixed point free, and the quotient space $V_A = (\mathbb{C}^n \setminus \{0\})/G_A$ is called a (linear) *Hopf manifold*. It is diffeomorphic to $S^1 \times S^{2n-1}$.

Proposition. *For a Hopf manifold* $\dim H^0(V_A, \Theta) = \dim H^1(V_A, \Theta)$, *and* $H^i(V_A, \Theta) = 0$ *for* $i > 1$. *These groups can be computed with the exact sequence*

$$0 \to H^0(V_A, \Theta) \longrightarrow H^0(\mathbb{C}^n, \Theta_n) \xrightarrow{1-A_*} H^0(\mathbb{C}^n, \Theta_n) \longrightarrow H^1(V_A, \Theta) \to 0 .$$

Proof [Do1]. Let $\mathcal{U}$ be a covering of V_A with simply connected open sets, which are Stein spaces. Let $\widetilde{\mathcal{U}}$ be the covering of $\mathbb{C}^n \setminus \{0\}$, consisting of $\pi^{-1}(U_i)$ for $U_i \in \mathcal{U}$, where $\pi: \mathbb{C}^n \setminus \{0\} \to V_A$ is the canonical projection. The matrix A acts on vector fields on $\mathbb{C}^n$, or subsets, by $(A_* \vartheta)(z) = A \cdot \vartheta(A^{-1}z)$. The covering $\widetilde{\mathcal{U}}$ consists of Stein open sets, invariant under A. The sequence of cochain complexes

$$0 \longrightarrow C^*(\mathcal{U}, \Theta) \longrightarrow C^*(\widetilde{\mathcal{U}}, \Theta_n) \xrightarrow{1-A_*} C^*(\widetilde{\mathcal{U}}, \Theta_n) \longrightarrow 0$$

is exact. Only the surjectivity on the right has to be proved, and this can be done for each open set separately. Let $V = U_{i_0 \dots i_p}$ be one of those, and choose a $\widetilde{V}_0 \subset \widetilde{V}$, which is isomorphic to V. Then $\widetilde{V} = \cup_{k \in \mathbb{Z}} \widetilde{V}_k := \cup_{k \in \mathbb{Z}} A^k(\widetilde{V}_0)$. Let $\vartheta = (\vartheta_k)_{k \in \mathbb{Z}} \in \Gamma(\widetilde{V}, \Theta_n)$. Define inductively ψ by $\psi_0 = \vartheta_0$, $\psi_k = \vartheta_k + A_* \psi_{k-1}$ for $k > 0$, and $\psi_k = A_*^{-1}(\psi_{k+1} - \vartheta_{k+1})$ for $k < 0$. Then $(1 - A_*)\psi = \vartheta$.

The short exact sequence induces a long exact cohomology sequence. As $H^i(\mathbb{C}^n \setminus \{0\}, \Theta_n) = \oplus_k H^i(\mathbb{C}^n \setminus \{0\}, \mathcal{O}_n)\partial/\partial z_k$, this group vanishes for $0 < i < n$. A direct computation with a covering with n open sets shows that the map $1 - A_*: H^n(\mathbb{C}^n \setminus \{0\}, \Theta_n) \longrightarrow H^n(\mathbb{C}^n \setminus \{0\}, \Theta_n)$ is an isomorphism. As every section of $H^0(\mathbb{C}^n \setminus \{0\}, \Theta_n)$ extends to $\mathbb{C}^n$, we get the sequence of the statement. $\square$

We now specialise to the case of surfaces (for the general case cf. [Pa3]), and we take coordinates (x, y) on $\mathbb{C}^2$. Let A be a diagonal matrix with two distinct eigenvalues λ and μ. One computes $(1 - A_*)x^i y^j \partial/\partial x = \lambda^{1-i}\mu^{-j}x^i y^j \partial/\partial x$. We have $\lambda^{1-i} \neq \mu^j$, if $i > 1$, because $|\lambda| < 1$ and $|\mu| < 1$. We get equality for $(i, j) = (1, 0)$, and also for $i = 0$, if the eigenvalues satisfy the resonance condition $\lambda = \mu^j$ for some j. These calculations lead to the following results, see also [We]:

1. If $A = \begin{pmatrix} \lambda & 0 \\ 0 & \lambda \end{pmatrix}$, then $h^1(V_A, \Theta) = 4$, and versal deformation is given by changing the matrix A into $A + M$, with M a (2×2) matrix from a suitable neighbourhood of $0 \in M(2, 2; \mathbb{C})$.

2. If $A = \begin{pmatrix} \lambda & 1 \\ 0 & \lambda \end{pmatrix}$, then $h^1(V_A, \Theta) = 2$, and the versal family is built with matrices
$$\begin{pmatrix} \lambda + t_1 & 1 \\ t_2 & \lambda + t_1 \end{pmatrix} .$$

3. If the eigenvalues of A are resonant, so $A = \begin{pmatrix} \lambda & 0 \\ 0 & \lambda^d \end{pmatrix}$ for some d, then $h^1(V_A, \Theta) = 3$, and among the fibres of the versal deformation are also nonlinear Hopf surfaces, given as $\mathbb{C}^2/\langle f \rangle$ for a non-linear contraction. The versal family is determined by
$$f(x, y; t_1, t_2, s) = \big((\lambda + t_1)x, (\lambda^d + t_2)y + sx^d\big) .$$

4. For $f(x, y) = (\lambda x, \lambda^d y + sx^d)$, the versal family is two-dimensional, given by the formula of (3.) with $s \equiv 1$.

5. For the general non-resonant matrix we have $h^1(V_A, \Theta) = 2$, and in normal form the family consists of diagonal matrices.

Example: non-vanishing obstructions [Do2]. Let X be a complex torus $\mathbb{C}^2/\Gamma$ and consider the product manifold $M = X \times \mathbb{P}^1$. The tangent bundle splits: $\Theta_M = p_1^* \Theta_X \oplus p_2^* \Theta_{\mathbb{P}^1}$. From the Leray spectral sequences for the projections one obtains
$$H^1(M, p_1^* \Theta_X) = H^0(\mathbb{P}^1, \mathcal{O}_{\mathbb{P}^1}) \otimes H^1(X, \Theta_X) \cong H^1(X, \Theta_X) ,$$
and
$$H^1(M, p_2^* \Theta_{\mathbb{P}^1}) = H^1(X, \mathcal{O}_X) \otimes H^0(\mathbb{P}^1, \Theta_{\mathbb{P}^1}) \cong \mathbb{C}^2 \otimes \mathfrak{sl}_2 .$$

The first group corresponds to deformations of X. The behaviour of the deformations from $H^1(p_2^* \Theta_{\mathbb{P}^1})$ does not depend on the particular value of the moduli of X, so we fix X and Γ. The Lie bracket on $H^*(M, p_2^* \Theta_{\mathbb{P}^1})$ is a combination of the product on $H^*(\mathcal{O}_X)$ and the Lie product of global vector fields on $\mathbb{P}^1$:
$$H^1(p_2^* \Theta_{\mathbb{P}^1}) \otimes H^1(p_2^* \Theta_{\mathbb{P}^1}) \xrightarrow{[\cdot, \cdot]} H^2(p_2^* \Theta_{\mathbb{P}^1}):$$
$$[f_1 \otimes \vartheta_1, f_2 \otimes \vartheta_2] = (f_1 \wedge f_2) \otimes [\vartheta_1, \vartheta_2] .$$

The obstruction $[\phi, \phi]$ vanishes for an $\phi \in H^1(X, \mathcal{O}_X) \otimes H^0(\mathbb{P}^1, \Theta_{\mathbb{P}^1})$ if and only if ϕ is a pure tensor: $\phi = f \otimes \vartheta$.

The manifold M is a trivial $\mathbb{P}^1$-bundle over X. We can describe deformations of this bundle by deformations of the trivial representation $\pi_1(X) = \Gamma \to PGL_2$. As Γ is abelian, the image must be contained in a 1-parameter subgroup $\exp(t\vartheta)$ for some $\vartheta \in \mathfrak{sl}_2$. The group $H^1(\mathcal{O}_X) \otimes H^0(\Theta_{\mathbb{P}^1})$ can be identified with $\operatorname{Hom}(\Gamma, \mathfrak{sl}_2)/\operatorname{Hom}_{\mathbb{C}}(\mathbb{C}^2, \mathfrak{sl}_2)$. A pure tensor corresponds to a homomorphism $\rho \colon \Gamma \to \mathfrak{sl}_2$, whose image is contained in a 1-dimensional linear subspace. The deformed manifold is the quotient of $\mathbb{C}^2 \times \mathbb{P}^1$ under the action

$$\gamma \cdot (x, y) = (x + \gamma, \exp(\rho(\gamma)) \cdot y)$$

of Γ. This construction defines a family with as base space the set of pure tensors, which is isomorphic to the cone over the Segre embedding of $\mathbb{P}^1 \times \mathbb{P}^2$.

8 How to solve the deformation equation

In the previous Chapter we have seen that deformations of compact complex manifolds are solutions of the equation

$$\bar{\partial}\,\vartheta + \tfrac{1}{2}[\vartheta,\vartheta] = 0$$

in the Dolbeault complex, which computes $H^1(M,\Theta)$. Also other deformation problems could be described by a similar deformation equation. The general situation is following: the problem is governed by a complex with a composition product, which makes the cohomology of the complex into a graded Lie algebra. An introduction to this point of view is given in [Nij]; the example, treated in detail there, is the deformation theory of associative algebras, due to GERSTENHABER [Ge], see also Chap. 5. LAUDAL writes: 'It is now folklore that the hull of a deformation functor of an algebraic geometric object, in some way is determined by the appropriate cohomology of the object and its "Massey products".' [La] A recent paper in this direction is [GM].

We will now discuss methods to solve the deformation equation in the abstract set-up of a complex $K^\bullet$ with a bracket operation, which descends to the cohomology. We think of the complex $K^\bullet$ as describing a deformation problem, with first order infinitesimal deformations given by $H^1(K^\bullet)$ and obstructions lying in $H^2(K^\bullet)$, while $H^0(K^\bullet)$ gives infinitesimal automorphisms. For $\varphi \in K^1$ we consider the equation

$$d\,\varphi + \tfrac{1}{2}[\varphi,\varphi] = 0 . \tag{8.1}$$

The first method to solve this equation is with a power series Ansatz. We write

$$\varphi = t\varphi_1 + t^2\varphi_2 + t^3\varphi_3 + \cdots$$

and substitute in (8.1). In this formula t is a parameter, which we use in a naive sense. Collecting powers of t we find the equations

$$0 = d\,\varphi_1$$

$$0 = d\,\varphi_2 + \tfrac{1}{2}[\varphi_1,\varphi_1]$$

$$0 = d\,\varphi_3 + [\varphi_1,\varphi_2]$$

$$\vdots$$

$$0 = d\,\varphi_n + \tfrac{1}{2}\sum_{i=1}^{n-1}[\varphi_i,\varphi_{n-i}] .$$

The first equation states that φ_1 is a cocycle, in accordance with the fact that the equivalence classes of first order infinitesimal deformations are given by $H^1(K^\bullet)$. The second equation gives the primary obstruction: the condition for extending φ_1 is that the cocycle $[\varphi_1, \varphi_1]$ is a coboundary; in other words, if the cohomology class of $[\varphi_1, \varphi_1]$ in $H^2(K^\bullet)$ is zero, one can find a φ_2 which is determined up to cocycles. The secondary obstruction is only defined, if φ_2 can be found; we can still change the specific choice of φ_2, giving an indeterminacy characteristic of Massey triple products.

This procedure tries to find a curve in the solution space, and the higher-order obstructions depend on the choices made in earlier steps. Instead we shall try to find the 'general' solution by a multi-variable power series Ansatz. We should clarify the meaning of 'general' solution. The best way to do that uses the categorical language of formal deformation theory [GM]. We give now the definitions, but will not pursue this further. Let $C^1(K^\bullet)$ be a complement to the 1-coboundaries $B^1(K^\bullet) \subset K^1$, and define the functor F_K from the category $\mathcal{C}$ of Artinian algebras to (Sets) by

$$F_K(A) = \{\, \varphi \in C^1(K^\bullet) \otimes \mathfrak{m}_A \mid d\varphi + \tfrac{1}{2}[\varphi, \varphi] = 0 \,\} \, .$$

If $\dim H^1(K^\bullet) < \infty$, this functor has a hull, because the natural map

$$F_K(A' \times_A A'') \longrightarrow F_K(A') \times_{F_K(A)} F_K(A'')$$

is bijective for any triple $A' \to A \leftarrow A''$ of Artinian local algebras. We note that $C^1(K^\bullet)$ is in general infinite dimensional. A 'general' solution is now nothing else than a miniversal object.

We come back to our multi-variable Ansatz. Let $\dim H^1(K^\bullet) = \tau$, and choose representatives $\varphi_1, \ldots, \varphi_\tau \in C^1(K^\bullet)$ of a basis. Let $t = (t_1, \ldots, t_\tau)$ be the corresponding coordinates. We construct the local ring S of the solution space as quotient of $\mathbb{C}[[t]]$; let $\mathfrak{m}_\tau$ be its maximal ideal. Over $S_1 := \mathbb{C}[[t]]/\mathfrak{m}_\tau^2$ we have the solution $\sum t_i \varphi_i$. To find the higher order terms we write

$$\varphi = \sum_{|\alpha| > 1} t^\alpha \varphi_\alpha \, ,$$

where this time we use a multi-variable power series, and multi-index notation. The primary obstruction comes from

$$\sum_{|\alpha|=2} t^\alpha \, d\varphi_\alpha + \frac{1}{2} \sum_{|i|=|j|=1} t^i t^j \, [\varphi_i, \varphi_j] = 0 \, . \tag{8.2}$$

We can express the class of $[\varphi_i, \varphi_j]$ in $H^2(K^\bullet)$ in terms of a basis $\Omega_1, \ldots, \Omega_s$ as $cl([\varphi_i, \varphi_j]) = \sum_k c_{ij}^k \Omega_k$. The equation (8.2) is solvable, if and only if

$$g_2^{(k)} := \frac{1}{2} \sum_{|i|=|j|=1} c_{ij}^k \, t^i t^j = 0 \, , \qquad \text{for all } k.$$

It is possible that some (or all) $g_2^{(k)}$ are zero, even if $\dim H^2(K^\bullet) > 0$. Set $S_2 := \mathbb{C}[[t]]/(g_2) + \mathfrak{m}_\tau^3$, and choose a basis B_2 of monomials for $\mathfrak{m}_\tau^2/(g_2) + \mathfrak{m}_\tau^3$; this can be done with a standard basis of the ideal (g_2). We will denote the set of exponents of these monomials also by B_2. Over S_2 we can solve (8.2): there are $\phi_\alpha \in C^1(K^\bullet)$, with $\alpha \in B_2$, such that

$$\sum_{\alpha \in B2} t^\alpha \, d\varphi_\alpha + \tfrac{1}{2} \sum_{|i|=|j|=1} t^i t^j [\varphi_i, \varphi_j] \equiv 0 \pmod{g_2} .$$

The φ_α are not unique, but determined up to elements $\psi_\alpha \in C^1(K^\bullet)$ with $d\psi_\alpha = 0$, or in other words the possible lifts form a homogeneous space under $H^1(K^\bullet)$. For the next step we have to solve the equation

$$\sum_{|\alpha|=3} t^\alpha \, d\varphi_\alpha + \sum_{\substack{|i|=1 \\ \alpha \in B_2}} t^i t^\alpha [\varphi_i, \varphi_\alpha] \equiv 0 \pmod{g_2} . \qquad (8.3)$$

Note that although the ideal (g_2) is defined in $\mathbb{C}[[t]]/\mathfrak{m}_\tau^3$, the ideal $\mathfrak{m}_\tau(g_2) \subset \mathfrak{m}_\tau^3/\mathfrak{m}_\tau^4$ is well-defined and does not depend on the extension of (g_2) to $\mathbb{C}[[t]]/\mathfrak{m}_\tau^4$. The class of $[\varphi_i, \varphi_\alpha]$ in $H^2(K^\bullet)$ is a *Massey triple product*. Write $cl([\varphi_i, \varphi_\alpha]) = \sum_k c_{i\alpha}^k \Omega_k$. This gives

$$g_3^{(k)} = \sum_{\substack{|i|=1 \\ \alpha \in B_2}} c_{i\alpha}^k t^i t^\alpha ,$$

which defines the extension of (g_2). Set $G_3^{(k)} = g_2^{(k)} + g_3^{(k)}$, and define S_3 as $\mathbb{C}[[t]]/(G_3) + \mathfrak{m}_\tau^4$. Choose a basis B_3 of monomials for $\mathfrak{m}_\tau^3 \cap (\mathbb{C}[[t]]/(G_3) + \mathfrak{m}_\tau^4)$. Over S_3 we can solve (8.3) using φ_α with $\alpha \in B_3$.

We will not continue, as the notation becomes cumbersome and anyhow the general procedure should be clear by now (see also [La]). The number of non-trivial equations $G_n^{(k)}$ becomes stationary. The problem with a power series Ansatz is that the process may never end. The computation will however always be finite, if our problem is graded, and we only consider deformations of negative degree. The convention is here that a deformation corresponds to a $\partial/\partial t_i$, so the parameters t_i have positive weight, and therefore we have a bound on the possible exponents α.

The second method to solve the deformation equation uses the implicit function theorem — it was used by KURANISHI in the case of compact complex manifolds. The vector spaces K^i tend to be infinite dimensional, so we have to specify some analytic structure in which the implicit function theorem holds. The necessity to shrink coverings leads to additional problems. To overcome these, PALAMODOV introduced the category of *PO*-spaces [Pal]. The most general results are in the work of BINGENER [Bi2].

To simplify matters, we assume that the K^i are complex Banach spaces. Furthermore we assume the existence of a splitting of the sequences

$$0 \longrightarrow Z^i(K^\bullet) \longrightarrow K^i \overset{d}{\longrightarrow} B^{i+1}(K^\bullet) \longrightarrow 0$$

and

$$0 \longrightarrow B^i(K^\bullet) \longrightarrow Z^i(K^\bullet) \longrightarrow H^i(K^\bullet) \longrightarrow 0 \,.$$

Let A^i be the image of the splitting $B^{i+1}(K^\bullet) \to K^i$, and H^i the image of $H^i(K^\bullet) \to Z^i(K^\bullet)$, and write also Z^i, B^i for $Z^i(K^\bullet)$ and $B^i(K^\bullet)$. We have the decomposition

$$K^i = Z^i + A^i = B^i + H^i + A^i.$$

Denote the corresponding projection maps by π_B, π_H and π_A. The hard part in the proofs of the existence of versal deformations is to establish such decompositions. In the case of compact complex manifolds they follow from the Hodge decomposition [Ku].

The goal is to show that the set

$$\{ \varphi \in H^1 + A^1 \mid d\varphi + \tfrac{1}{2}[\varphi, \varphi] = 0 \}$$

has the structure of a finite dimensional complex space. For this one splits the equation (8.1) in three equations:

$$d\varphi + \tfrac{1}{2}\pi_B[\varphi, \varphi] = 0 \,, \qquad\qquad (a)$$
$$\pi_H[\varphi, \varphi] = 0 \,, \qquad\qquad (b)$$
$$\pi_A[\varphi, \varphi] = 0 \,. \qquad\qquad (c)$$

The left-hand side of equation (a) defines a map $D\colon K^1 \to B^2$ with derivative d at the origin, so by the implicit function theorem there is an isomorphism F from a neighbourhood U of the origin in Z^1 onto a neighbourhood of the origin in $D^{-1}(0)$. We now consider the equation (b) and define

$$S = \{ \varphi \in H^1 \cap U \mid \pi_H[F(\varphi), F(\varphi)] = 0 \} \,.$$

The space S is contained in the finite-dimensional vector space H^1, and this gives the complex structure we are after. The third equation (c) gives no further conditions, because for small φ it follows from the first two. To see this, we remember that d maps A^2 isomorphically to B^3, and compute $d\pi_A[\varphi, \varphi] = d[\varphi, \varphi] = 2[d\varphi, \varphi]$ by the compatibility of d and the bracket. By (a) this again is equal to $-[\pi_B[\varphi, \varphi], \varphi] = [\pi_A[\varphi, \varphi], \varphi]$, where we use (b) and the decomposition $1 = \pi_B + \pi_H + \pi_A$. As $d|_{A^2}$ is invertible, we have, writing ψ for $\pi_A[\varphi, \varphi]$, the equation $\psi = (d|_{A^2})^{-1}[\psi, \varphi]$ and by continuity there is a constant C, independent of ψ and φ, such that

$$||\psi|| \le C||\psi||\,||\varphi|| \,.$$

Therefore, for $||\varphi||$ small enough we have $C||\varphi|| < 1$, showing that $\psi = \pi_A[\varphi, \varphi] = 0$.

Note that again we have constructed the solution space as fibre of a map $H^1 \to H^2$.

9 Convergence for isolated singularities

In this section we give the easy part of GRAUERT's proof [Gra] of the existence of versal deformations for isolated singularities; we follow [GH]. The actual work for showing convergence is in the following *Grauert approximation theorem*, which GALLIGO and HOUZEL attribute in this form to VERDIER.

Consider four sets of variables, being $x = (x_1, \ldots, x_n)$, $u = (u_1, \ldots, u_q)$, $\Phi = (\Phi_1, \ldots, \Phi_m)$ and $\Psi = (\Psi_1, \ldots, \Psi_p)$. Let I be an ideal in $\mathbb{C}\{x\}$, and $\mathfrak{m}$ the maximal ideal in $\mathbb{C}\{x\}$. A *solution of the system of equations* $R \equiv 0$ (mod I), where $R \in \mathbb{C}\{x, u, \Phi, \Psi\}^l$, is a pair (ϕ, ψ) with $\phi \in \mathbb{C}\{x\}$ and $\psi \in \mathbb{C}\{x, u\}$, such that

$$R(x, u, \phi(x), \psi(x, u)) \equiv 0 \pmod{I} .$$

A *solution up to order* i consists of a $\phi \in \mathbb{C}[x]$ and $\psi \in \mathbb{C}\{u\}[x]$, with degree in x at most i, such that

$$R(x, u, \phi(x), \psi(x, u)) \equiv 0 \pmod{I + \mathfrak{m}^{i+1}} .$$

Theorem. *Let $R \equiv 0$ (mod I) be a system of equations as above, and suppose it has a solution up to order i_0. If every solution (ϕ_i, ψ_i) up to order i, $i \geq i_0$, extends to a solution $(\phi_i + \delta_i, \psi_i + \gamma_i)$ up to order $i+1$, with $\delta_i \in \mathbb{C}[x]$ and $\gamma_i \in \mathbb{C}\{u\}[x]$ homogeneous polynomials in x of degree $i + 1$, then the system has an analytic solution.*

For the proof we refer to [Gal]. The assumptions for the theorem are much stronger than for Artin approximation, but the conclusion we want, namely that ϕ is independent of u, is also stronger. An example of GABRIELOV shows that in general a formal solution $\widehat{\phi}(x)$ does not necessarily lead to an analytic solution, independent of u [Gab].

Suppose X_0 is a singularity with $\tau := \dim T^1 < \infty$. By the results of SCHLESSINGER (see Chap. 6) a formal semi-universal deformation exists. We describe it with 'explicit' equations: the base space $\overline{S} = \varprojlim S_m$ is defined by the ideal $(\overline{g}) := (\overline{g}_1, \ldots, \overline{g}_p)$ in $\mathbb{C}[[s]] = \mathbb{C}[[s_1, \ldots, s_\tau]]$, and S_m is defined by $(\overline{g}) + \mathfrak{m}^{m+1}$. From now on we consider $\overline{g}$ as a row vector, consisting of a set of generators of the ideal, and we do not refer to entries of this vector separately. The vector $g_m \in (\mathbb{C}[s])^p$ of polynomials with degree at most m is the reduction modulo $\mathfrak{m}^{m+1}$ of $\overline{g}$: $g_m \equiv \overline{g}$ (mod $\mathfrak{m}^{m+1}$). The ideal $(\overline{F}) \subset$

$\mathbb{C}\{x\}[[s]]/(g)\mathbb{C}\{x\}[[s]]$ describes the total space $\overline{X} = \varprojlim X_m$. This does not mean that $\overline{F}$ defines a deformation of X_0 over $\overline{S}$. There exist matrices $\overline{R}$ and $\overline{Q}$ with entries in $\mathbb{C}\{x\}[[s]]$, such that $\overline{F}\cdot\overline{R} = \overline{g}\cdot\overline{Q}$: the vector $F_m = \overline{F}$ mod $\mathfrak{m}_\tau^{m+1}$ defines a deformation over S_m, so the relations $r = R_0$ between $f = F_0$ can be lifted: $F_m \cdot R_m \equiv 0 \pmod{(g) + \mathfrak{m}^{m+1}}$, or more explicitly:

$$F_m \cdot R_m \equiv g_m \cdot Q_m \pmod{\mathfrak{m}_\tau^{m+1}}.$$

In the preceding equations the ideal $\mathfrak{m}_\tau$ is the ideal in $\mathbb{C}\{x\}[[s]]$, generated by the maximal ideal of $\mathbb{C}[[s]]$.

Now consider the system of equations:

$$\mathcal{S}: \quad \begin{cases} g & \equiv & g_m \bmod \mathfrak{m}_\tau^{m+1} \\ F & \equiv & F_m \bmod \mathfrak{m}_\tau^{m+1} \\ R & \equiv & R_m \bmod \mathfrak{m}_\tau^{m+1} \\ F \cdot R & = & g \cdot Q. \end{cases}$$

We will apply Grauert's approximation theorem to this system with $i_0 = m$ for a suitably chosen m. As preparation we need GRAUERT's theory of extensions [Gra, I.4].

Definition. Let $\mathcal{H}_m = \mathbb{C}[[s]]/\mathfrak{m}^{m+1}$ be the Artinian algebra of polynomials of degree at most m in s. An ideal $J_{m+i} \subset \mathcal{H}_{m+i}$ (where $i \geq 1$) is an *extension* of $J_m \subset \mathcal{H}_m$, if the image of J_{m+i} under the natural projection $\rho: \mathcal{H}_{m+i} \to \mathcal{H}_m$ equals J_m. An extension J_{m+i} is a *minimal extension*, if $J'_{m+i} = J_{m+i}$ for every other extension $J'_{m+i} \subset J_{m+i}$.

Lemma. *Let $J_m \subset \mathcal{H}_m$ be minimally generated by $g = (g_1, \ldots, g_p)$. An extension J_{m+1} of J_m is minimal if and only if J_{m+1} can be generated by $G = (G_1, \ldots, G_p)$ with $\rho(G) = g$. Furthermore, if J_{m+1} and J'_{m+1} are minimal extensions of J_m, then $J_{m+1} \cap \mathfrak{m}^{m+1}/\mathfrak{m}^{m+2} = J'_{m+1} \cap \mathfrak{m}^{m+1}/\mathfrak{m}^{m+2}$.*

Proof. Let J_{m+1} be a minimal extension, and choose a lift $G \subset (J_{m+1})^{\oplus p}$ of g. By minimality the ideal generated by (G) equals J_{m+1}. Conversely, if $J' \subset J_{m+1} = (G)$ is an extension, then $(J')^{\oplus p}$ contains a lift G' of g, and as $J' \subset (G)$, there exists a matrix A with $G' = GA$, and $\rho(A) = \mathrm{Id}$. Therefore the matrix A is invertible, and $J' = (G)$.

Let G and G' be the lifts of g to J_{m+1} and J'_{m+1}. Let $h = \sum a_i G_i \in J_{m+1} \cap \mathfrak{m}^{m+1}$. As $\rho(h) = 0$, we have $\sum \rho(a_i)g_i = 0$; because (g) is a minimal set of generators, one has $0 = (\rho(a_i))(0) = a_i(0)$ for all i, which implies that $a_i \in \mathfrak{m}$. But then $\sum a_i G_i = \sum a_i g_i = \sum a_i G'_i$. $\qquad\square$

If J is an ideal in $\mathbb{C}[[s]]$, then there exists an index m, such that for $q \geq m$ the sequence $J_q = (J + \mathfrak{m}^{q+1})/\mathfrak{m}^{q+1}$ is a sequence of minimal extensions. Similar results hold for submodules of the free modules $\mathcal{H}_m^{\oplus p}$.

Theorem. *Let X_0 be a singularity with $\tau = \dim T^1 < \infty$. There exists an analytic deformation $X \to S$ of X_0, which is formally miniversal.*

Proof. Consider the formal, formally miniversal deformation, described above. Let $J = (\bar{g}) \subset \mathbb{C}[[s]]$ be the ideal of $\overline{S}$, and choose m such that both the sequence J_q, $q \geq m$, and the sequence of relation modules $\mathcal{R}_q$ between the generators consists of minimal extensions. Note that the minimal number of generators of J_q is constant. For each $q \geq m$ the system S has a solution (g_q, F_q, R_q, Q_q) up to order q.

Let (g'_q, F'_q, R'_q, Q'_q) be any solution of S up to order q. It defines a deformation X'_q of X_0 over $S'_q = \mathrm{Specan}\left(\mathcal{H}_q/(g'_q)\right)$, and the extensions $(g'_m), \ldots,$ (g'_q) are minimal. As $\overline{S}$ is miniversal, we have a diagram

$$
\begin{array}{ccc}
X'_q & \longrightarrow & X_q \\
\downarrow & & \downarrow \\
S'_q & \longrightarrow & S_q
\end{array}
$$

where the horizontal maps are the identity up to order m. The map $\mathcal{H}_q/(g_q) \to \mathcal{H}_q/(g'_q)$ is surjective, because it is the identity up to order $m \geq 1$. The dimensions as $\mathbb{C}$-vector space of these algebras is equal, because all extensions are minimal. Therefore $S'_q \to S_q$ is an isomorphism, which is induced by an automorphism ϕ_q of $\mathcal{H}_q$, and X'_q is isomorphic to X_q via an automorphism Φ_q of $\mathbb{C}\{x\}[s]/\mathfrak{m}_T^{q+1}$ lying over ϕ_q. Lift ϕ_q to an automorphism ϕ_{q+1} of $\mathcal{H}_{q+1}$ and Φ_q to an automorphism Φ_{q+1} of $\mathbb{C}\{x\}[s]/\mathfrak{m}_T^{q+2}$. Set $(g'_{q+1}, F'_{q+1}) = (\phi_{q+1}(g_{q+1}), \Phi_{q+1}(F_{q+1}))$. This defines a deformation, which extends $X'_q \to S'_q$.

By flatness the relations $F'_q R'_q \equiv 0 \pmod{(g'_q)}$ lift to relations

$$
F'_{q+1} R'_{q+1} \equiv 0 \pmod{(g'_{q+1}), \mathfrak{m}_T^{q+2}} \,,
$$

so $F'_{q+1} R'_{q+1} \equiv g'_{q+1} Q''_{q+1} \pmod{\mathfrak{m}_T^{q+2}}$ for some Q''_{q+1}. Thus $g'_q(\rho(Q''_{q+1}) - Q'_q) \equiv 0 \pmod{\mathfrak{m}_T^{q+1}}$, which is a relation between the g'_q up to order q, giving one between the g_q via our isomorphisms. By our choice of m this relation can be extended to order $q + 1$, yielding $g'_{q+1} \widetilde{Q}_{q+1} \equiv 0 \pmod{\mathfrak{m}_T^{q+2}}$. We set $Q'_{q+1} = Q''_{q+1} - \widetilde{Q}_{q+1}$. Then $\rho(Q'_{q+1}) = Q'_q$ and $F'_{q+1} R'_{q+1} \equiv g'_{q+1} Q'_{q+1} \pmod{\mathfrak{m}_T^{q+2}}$. The hypotheses of Grauert's approximation theorem are satisfied. $\qquad\square$

Theorem. *The deformation, constructed in the previous theorem, is (analytically) miniversal.*

Proof. We only write down the relevant equations. Let $f' \in (\mathbb{C}\{x,t\})^{\oplus k}$ define a deformation over T, with $\mathcal{O}_T = \mathbb{C}\{t\}/J'$, $t = (t_1, \ldots, t_r)$; we may suppose that $f'_0 = f_0$. As $X \to S$ is formally versal, there exist:

— r power series $h_i \in \mathbb{C}[[t]]$, inducing $h^*\colon \mathbb{C}[[s]] \to \mathbb{C}[[t]]$ with $h^*(J) \subset J'$,
— an automorphism Φ of $\mathbb{C}\{x\}[[t]]$, determined by the power series $\Phi(x_i)$,
— a $k \times k$-matrix M with entries in $\mathbb{C}\{x\}[[t]]$, such that $M_0 = \mathrm{Id}$,

satisfying the system of equations:

$$\mathcal{T}: \quad \begin{cases} f'(\Phi(x),t)M(x,t) \equiv f(x,h^*s) \pmod{J'} \\ \qquad\qquad g(h^*s) \in J' \ . \end{cases}$$

Moreover, if on a subspace $T' \subset T$ an analytic map $T' \to S$ is already given, we can take the above data compatible with it. The Grauert approximation theorem again gives an analytic solution. $\square$

10 Quotient singularities

The module T^1 can be defined as a cohomology group of a certain complex, and there are several ways to compute it, without using the equations of the singularity. ALTMANN starts from the description $T_X^1 = \mathrm{Ext}(\Omega_X^1, \mathcal{O}_X)$ to compute the infinitesimal deformations of affine toric varieties [Alt1]. For curve singularities one can compute on the normalisation, see [RV]. Most computations however are based on SCHLESSINGER's description [Schl2].

We recall the definition of T_X^1 for a singularity $(X,0) \subset (\mathbb{C}^e, 0)$ from Chap. 3:

$$0 \longrightarrow \Theta_X \longrightarrow \Theta_{\mathbb{C}^e}|_X \longrightarrow N_X \longrightarrow T_X^1 \longrightarrow 0. \tag{10.1}$$

Here $N_X = \mathrm{Hom}(I/I^2, \mathcal{O}_X)$ is the normal sheaf; the tangent sheaf of X is also defined as a dual: $\Theta_X = \mathrm{Hom}(\Omega_X^1, \mathcal{O}_X)$. These sheaves are therefore torsion free for reduced X.

Let $Z \subset X$ be a closed subset, containing the singular locus $\mathrm{Sing}\, X$ of X, and suppose that $\mathrm{depth}_Z X \geq 2$ (this condition is satisfied if X is normal and $Z = \mathrm{Sing}\, X$); put $U = X \setminus Z$. Then $H^0(U, \Theta_N|_X) = H^0(X, \Theta_N|_X)$, and similarly for N_X. This gives us

$$T_X^1 = \mathrm{coker}\{H^0(U, \Theta_N|_X) \to H^0(U, N_X)\}.$$

On the smooth space U the sheaf T^1 is zero, so the normal bundle sequence (10.1) is a three term short exact sequence, and the map defining T_X^1 occurs in its long exact cohomology sequence. Therefore we have the alternative description

$$T_X^1 = \ker\{H^1(U, \Theta_U) \to H^1(U, \Theta_N|_X)\}.$$

For $T_X^2 = \mathrm{coker}\{\mathrm{Hom}((\mathcal{O}_X)^{\oplus k}, \mathcal{O}_X) \to \mathrm{Hom}(R/R_0, \mathcal{O}_X)\}$ something similar can be done. We consider on X the exact sequence

$$0 \to R_X \to (\mathcal{O}_X)^{\oplus k} \to I/I^2 \to 0.$$

The kernel of the surjection $R/R_0 \to R_X$ is a torsion module, if X is reduced. Therefore we get as alternative description

$$0 \to N_X \longrightarrow \mathrm{Hom}((\mathcal{O}_X)^{\oplus k}, \mathcal{O}_X) \longrightarrow \mathrm{Hom}(R_X, \mathcal{O}_X) \longrightarrow T_X^2 \to 0.$$

Suppose the sheaf $\mathcal{T}^2$ has support contained in Z_2 with $\mathrm{depth}_{Z_2} X \geq 2$, and let $U_2 = X \setminus Z_2$. Then

$$T_X^2 = \ker\{H^1(U_2, N_X) \to H^1(U_2, (\mathcal{O}_n)^{\oplus k})\} .$$

Returning to T^1, if furthermore depth$_Z X \geq 3$, then $H^1(U, \Theta_N|_X) = 0$, because $\Theta_N|_X$ is a locally free sheaf. In that case $T_X^1 = H^1(U, \Theta_U)$. SCHLESSINGER uses this to prove the following result:

Theorem. *Quotient singularities, which are nonsingular in codimension two, are rigid.*

Let $X = \mathbb{C}^n/G$ with G a finite group, and let $V \subset \mathbb{C}^n$ be the locus where the action is fixed point free; then $U = V/G$, and $H^1(U, \Theta_U) = H^1(V, \Theta_n|V)^G$. Under the assumptions of the theorem $H^1(V, \Theta_n|V)^G = 0$. If the singular locus has codimension two, one can still compute $H^1(U, \Theta_U)$ as $H^1(V, \Theta_n|V)^G$, but now this group is not even finite dimensional; fortunately the same is true for $H^1(U, \Theta_N|_X)$. For two dimensional cyclic quotients T^1 has been computed in this way in [Pin2], and for the dihedral singularities in [BR]; for the remaining quotients see [BKR].

Example: two dimensional cyclic quotients. Let $X_{n,q} = \mathbb{C}^2/G_{n,q}$ be a cyclic quotient singularity, where $G_{n,q}$ is the subgroup of $Gl(2, \mathbb{C})$ generated by

$$\begin{pmatrix} \zeta_n & 0 \\ 0 & \zeta_n^q \end{pmatrix} ,$$

with ζ_n is a primitive n-th root of unity and $(n, q) = 1$. Its local ring is the ring $\mathbb{C}\{x, y\}^{G_{n,q}}$ of $G_{n,q}$-invariants. These singularities are the two-dimensional affine toric varieties. In that language we start with the lattice $N = \mathbb{Z}^2 + \mathbb{Z} \cdot \frac{1}{n}(1, q)$. The dual lattice M consist then just of the exponents of $G_{n,q}$-invariant Laurent monomials. As cone σ we take the first quadrant. The ring of invariants is then given by $\sigma^\vee \cap M$.

We recall the recursive definition of an (improper) continued fraction $[c_1, \ldots, c_r]$, which we consider as rational function of its entries: $[c_r] = c_r$, and

$$[c_i, c_{i+1}, \ldots, c_r] = c_i - \frac{1}{[c_{i+1}, \ldots, c_r]} .$$

One defines numbers $p_0, \ldots, p_{r+1}$ and $q_0, \ldots, q_{r+1}$ by

$$p_0 = 0 , \quad p_1 = 1 , \quad p_{i-1} + p_{i+1} = c_i p_i ,$$
$$q_{r+1} = 0 , \quad q_r = 1 , \quad q_{i+1} + q_{i-1} = c_i q_i .$$

With these numbers the partial continued fractions can be expressed as $[c_k, \ldots, c_r] = q_{k-1}/q_k$ and $[c_k, \ldots, c_1] = p_{k+1}/p_k$. A rational number t/s has a canonical continued fraction expansion with integers defined by: $c_1 = \lceil \frac{r}{s} \rceil$ and $[c_2, \ldots, c_r]$ is the continued fraction expansion of $s/(sc_1 - t)$.

One forms the continued fraction expansions

$$\frac{n}{q} = [b_1, \ldots, b_r] ,$$

$$\frac{n}{n - q} = [a_2, \ldots, a_{e-1}] .$$

The expansion of n/q gives the resolution: the exceptional divisor on the minimal resolution is a chain of smooth rational curves with self intersection $-b_i$. The expansion of $n/(n-q)$ is dual to that of n/q; one finds the one from the other with RIEMENSCHNEIDER's *point diagram* [Rie1]: place in the i-th row a_i-1 dots, the first one under the last one of the $(i-1)$-st row; the column j contains b_j-1 dots. We remark that $[b_1, \ldots, b_r, 1, a_{e-1}, \ldots, a_2] = 0$.

Example. Take $[a] = [2, 5, 2, 3]$, then $[b] = [3, 2, 2, 4, 2]$, and we have the diagram:

$$\begin{matrix} & \bullet & & & \\ \bullet & \bullet & \bullet & \bullet & \\ & & \bullet & & \\ & & \bullet & \bullet & \end{matrix}$$

The expansion of $n/(n-q)$ is important for the equations. We write i_ε and j_ε for the numbers p_ε and q_ε, associated with $[a]$. The invariant monomials

$$z_\varepsilon = x^{i_\varepsilon} y^{j_\varepsilon}, \qquad \varepsilon = 1, \ldots, e,$$

generate the ring of invariants. We obtain immediately equations $z_{\varepsilon-1} z_{\varepsilon+1} - z_\varepsilon^{a-\varepsilon}$. The number of generators of the ideal is the same as for the cone over the rational normal curve of degree $e-1$ and they can be conveniently written as quasi-determinantal (cf. p. 93):

$$\begin{vmatrix} z_1 & z_2 & \cdots & z_{e-2} & z_{e-1} \\ z_2 & z_3 & \cdots & z_{e-1} & z_e \\ z_2^{a_2-2} & & \cdots & & z_{e-1}^{a_{e-1}-2} \end{vmatrix}.$$

Write X for $X_{n,q}$ and $U = X \setminus \{0\}$. To compute

$$T_X^1 = \ker\{H^1(U, \Theta_U) \to H^1(U, \Theta_e|X)\},$$

we consider the covering $\mathcal{U}$ of U with two affine charts, $U_0 = \{z_1 \neq 0\}$, $U_{r+1} = \{z_e \neq 0\}$, which corresponds to the covering $U_0 = \{x \neq 0\}$, $U_{r+1} = \{y \neq 0\}$ of $\mathbb{C}^2 \setminus \{0\}$. Consider the Čech-complex:

$$\begin{array}{ccccccccc} 0 & \longrightarrow & C^0(\mathcal{U}, \Theta_X) & \longrightarrow & C^0(\mathcal{U}, \Theta_e|X) & \longrightarrow & C^0(\mathcal{U}, N_X) & \longrightarrow & 0 \\ & & \downarrow{\scriptstyle \delta} & & \downarrow{\scriptstyle \delta} & & \downarrow{\scriptstyle \delta} & & \\ 0 & \longrightarrow & C^1(\mathcal{U}, \Theta_X) & \longrightarrow & C^1(\mathcal{U}, \Theta_e|X) & \longrightarrow & C^1(\mathcal{U}, N_X) & \longrightarrow & 0 \end{array}$$

The torus action on X makes all modules bigraded. The elements of $C^1(\mathcal{U}, \Theta_X)$ in a fixed bidegree $(-a, -b)$ are of the form:

$$\vartheta = \frac{1}{x^a y^b}\left(\lambda x \frac{\partial}{\partial x} + \mu y \frac{\partial}{\partial y}\right).$$

Then $\vartheta(z_\varepsilon) = (i_\varepsilon \lambda + j_\varepsilon \mu) x^{i_\varepsilon - a} y^{j_\varepsilon - b}$, which is a coboundary if $a \leq i_\varepsilon$, $b \leq j_\varepsilon$ or $i_\varepsilon \lambda + j_\varepsilon \mu = 0$. If $x^a y^b$ is divisible by a product $z_\alpha z_\beta$ with $\alpha \neq \beta$, then the first two conditions are obviously not satisfied for $\varepsilon = \alpha$ and $\varepsilon = \beta$; as the sequence i_ε is increasing and j_ε decreasing, we obtain two independent equations for λ and μ. The result is:

Lemma. Let $X = X_{n,q}$ be a cyclic quotient with $\frac{n}{n-q} = [a_2, \ldots, a_{e-1}]$, and suppose $e > 3$. The dimension of T_X^1 is $(e - 4) + \sum_\varepsilon (a_\varepsilon - 1)$, and a basis is given by the vector fields:

$$\vartheta_\varepsilon^{(a)} = -\frac{1}{n} \frac{1}{(x^{i_\varepsilon} y^{j_\varepsilon})^a} \left(j_\varepsilon x \frac{\partial}{\partial x} - i_\varepsilon y \frac{\partial}{\partial y} \right), \qquad 1 \le a \le a_\varepsilon - 1,\ 2 \le \varepsilon \le e - 1,$$

$$\sigma_\varepsilon = -\frac{1}{n} \frac{1}{x^{i_\varepsilon} y^{j_\varepsilon}} \left((j_\varepsilon - j_{\varepsilon-1}) x \frac{\partial}{\partial x} - (i_\varepsilon - i_{\varepsilon-1}) y \frac{\partial}{\partial y} \right), \qquad 3 \le \varepsilon \le e - 2.$$

To compute the action on the equations, we express the vector fields in terms of the $\partial_\rho := \frac{\partial}{\partial z_\rho}$. From $x \frac{\partial}{\partial x} = \sum i_\rho z_\rho \partial_\rho$ and $y \frac{\partial}{\partial y} = \sum j_\rho z_\rho \partial_\rho$, we obtain

$$\vartheta_\varepsilon^{(a)} = -\frac{1}{n} \frac{1}{z_\varepsilon^a} \sum (i_\rho j_\varepsilon - i_\varepsilon j_\rho) z_\rho \partial_\rho,$$

$$\sigma_\varepsilon = -\frac{1}{n} \frac{1}{z_\varepsilon} \sum \left(i_\rho (j_\varepsilon - j_{\varepsilon-1}) - (i_\varepsilon - i_{\varepsilon-1}) j_\rho \right) z_\rho \partial_\rho.$$

With $a_\varepsilon i_\varepsilon = i_{\varepsilon-1} + i_{\varepsilon+1}$ one computes that $i_{\varepsilon-1} j_\varepsilon - i_\varepsilon j_{\varepsilon-1}$ is independent of ε, so equal to $i_0 j_1 - i_1 j_0 = n$. Therefore

$$\vartheta_\varepsilon^{(a)} = \frac{1}{z_\varepsilon^a} (\cdots - z_{\varepsilon-1} \partial_{\varepsilon-1} + z_{\varepsilon+1} \partial_{\varepsilon+1} + \cdots),$$

$$\sigma_\varepsilon = -\frac{1}{z_\varepsilon} (c \ldots + z_{\varepsilon-1} \partial_{\varepsilon-1} + z_\varepsilon \partial_\varepsilon + (a_\varepsilon - 1) z_{\varepsilon+1} \partial_{\varepsilon+1} + \cdots).$$

It is sufficient to give two coefficients c_i of the vector fields $\sum c_i z_i \partial_i$ above, because the c_i satisfy the equation $a_i c_i = c_{i-1} + c_{i+1}$.

ARNDT has proven that an infinitesimal deformation of X is completely determined by its action on the $e - 2$ equations $z_{\varepsilon-1} z_{\varepsilon+1} - z_\varepsilon^{a_\varepsilon}$ [Arn]. On the vector space T_X^1 we take coordinates $t_\varepsilon^{(a)}$, $1 \le a \le a_\varepsilon - 1$, $\varepsilon = 2, \ldots, e - 1$ and s_ε, $\varepsilon = 3, \ldots, e - 2$. We write the infinitesimal deformations as

$$z_{\varepsilon-1} (z_{\varepsilon+1} - s_{\varepsilon+1}) - (z_\varepsilon - s_\varepsilon) \left(z_\varepsilon^{a_\varepsilon - 1} + t_\varepsilon^{(1)} z_\varepsilon^{a_\varepsilon - 2} + \ldots + t_\varepsilon^{(a_\varepsilon - 1)} \right).$$

To make this formula valid for all ε we introduce dummy variables s_2, s_{e-1} and s_e, whose value we set equal to zero. If we lift the vector fields $\vartheta_\varepsilon^{(a)}$ and σ_ε to $C^0(N_X)$ by taking $\rho \ge \varepsilon$ on U_{r+1}, we obtain exactly

$$\vartheta_\varepsilon^{(a)} = \frac{\partial}{\partial t_\varepsilon^{(a)}}, \qquad \sigma_\varepsilon = \frac{\partial}{\partial s_\varepsilon}.$$

For $e = 3$ one has to modify the previous formulas. The versal deformation of A_{n-1} is $z_1 z_3 - (z_2^n + t^{(2)} z_2^{(n-2)} + \ldots + t^{(n-1)} z_2 + t^{(n)})$ and

$$\frac{\partial}{\partial t^{(a)}} = \vartheta^{(a)} = -\frac{1}{n} \frac{1}{(xy)^a} \left(x \frac{\partial}{\partial x} - y \frac{\partial}{\partial y} \right), \qquad 2 \le a \le n.$$

Obstructions. The question arises, if one can compute not only T^1, but also the first obstruction for a normal isolated singularity from a suitable covering $\mathcal{U}$ of $U = X \setminus \{0\}$. This was done by Michael JUNGE in his Hamburger Diplomarbeit [Ju], using the Kodaira-Spencer deformation theory of smooth manifolds.

We return to the the Čech-complex

$$
\begin{array}{ccccccccc}
0 & \longrightarrow & C^0(\mathcal{U}, \Theta_X) & \longrightarrow & C^0(\mathcal{U}, \Theta_e | X) & \longrightarrow & C^0(\mathcal{U}, N_X) & \longrightarrow & 0 \\
& & \downarrow{\scriptstyle \delta} & & \downarrow{\scriptstyle \delta} & & \downarrow{\scriptstyle \delta} & & \\
0 & \longrightarrow & C^1(\mathcal{U}, \Theta_X) & \longrightarrow & C^1(\mathcal{U}, \Theta_e | X) & \longrightarrow & C^1(\mathcal{U}, N_X) & \longrightarrow & 0
\end{array}
$$

With a little cheating it is not difficult to find the obstruction: let $[\vartheta] \in H^1(\mathcal{U}, \Theta_U)$ such that $[\vartheta] = 0$ in $H^1(\mathcal{U}, \Theta_e | X)$; then $\vartheta_{ij} = \varphi_j - \varphi_i$ with $\varphi_i \in \Gamma(U_i, \Theta_e | X)$. On U_i the infinitesimal deformation is given by $f + \varepsilon \varphi_i(f)$, and we lift the relations $fr = 0$ to

$$
0 = (1 + \varepsilon \varphi_i)(fr) \equiv \big(f + \varepsilon \varphi_i(f) \big) \big(r + \varepsilon \varphi_i(r) \big) \bmod \varepsilon^2 .
$$

The obstruction is then $\varphi_i(f)\varphi_i(r)$. From $\varphi_i\big(\varphi_i(fr)\big) = 0$ it follows that

$$
\varphi_i(f)\varphi_i(r) \equiv -\tfrac{1}{2}\varphi_i^2(f) r \quad (\bmod f) .
$$

In $T^2 = \ker\{H^1(N_U) \to H^1(\mathcal{O}_U^k)\}$ the obstruction is represented by $\tfrac{1}{2}\varphi_i^2(f) - \tfrac{1}{2}\varphi_j^2(f) = \tfrac{1}{2}[\varphi_i, \varphi_j](f)$ (because $\varphi_i(f) = \varphi_j(f)$), or in other words: the obstruction is the class $\tfrac{1}{2}[\varphi_i, \varphi_j] \in H^1(N_U)$.

Unfortunately one cannot compute in this way on U: the action of φ_i an elements of $\mathcal{O}_{U_i}$ is not well-defined, as $\varphi_i(0) = 0$, but $\varphi_i(f) \neq 0$. So one has to compute with functions, defined on open sets of $\mathbb{C}^e$ (as one does in practice). But then the equality $\varphi_i(f) = \varphi_j(f)$ holds only modulo (f).

Instead of $\Theta_e | X = i^* \Theta_e$ for $i \colon X \to \mathbb{C}^e$, we consider sections of $i^{-1}\Theta_e$ over U_i: this are germs of vector fields on $\mathbb{C}^e$ along U_i. Furthermore, for an ideal $I \in \mathcal{O}_e$ we define $\Theta_I = \{ \vartheta \in \Theta_e \mid \vartheta(I) \subset I \}$; let $I = I(X)$ be the ideal generated by (f). Then there is an exact sequence of Čech-complexes

$$
0 \longrightarrow C^*(\mathcal{U}, i^{-1}\Theta_I) \longrightarrow C^*(\mathcal{U}, i^{-1}\Theta_e) \longrightarrow C^*(\mathcal{U}, N_X) \longrightarrow 0 .
$$

If $\Gamma(U_i \cap U_j, i^{-1}\Theta_I) \ni \vartheta_{ij} = \varphi_j - \varphi_i$, then $\varphi_i(f)$ determines an element of $\mathcal{O}_X$, so there is a matrix $A \in \mathrm{Mat}_k\big(\Gamma(U_i, i^{-1}\Theta_e)\big)$ such that $\varphi_i(f) - fA$ is holomorphic.

Definition. Let $U \subset \mathbb{C}^e$ be an open subset. Let the Lie algebra $\mathcal{A}(U)$ be the semi-direct product of $\Theta(U)$ with $\mathrm{Mat}_k(\mathcal{O}(U))$: for $\Phi_i = (\varphi_i, A_i)$ with φ_i a vector field, we have

$$
[\Phi_1, \Phi_2] = \big([\varphi_1, \varphi_2], \varphi_1(A_2) - \varphi_2(A_1) + [A_1, A_2] \big) .
$$

Lemma (Leibniz rule). *Let $\Phi = (\varphi, A) \in \mathcal{A}(U)$ act on functions via φ, on row vectors as $\Phi(f) = \varphi(f) - fA$ and on column vectors as $\Phi(r) = \varphi(r) + Ar$. Then $\Phi(fr) = \Phi(f)r + f\Phi(r)$.*

We can do the same as above, but now with $\Phi_i \in \Gamma(U_i, i^{-1}\mathcal{A}(U))$. As $[\Phi_i, \Phi_j](f) \equiv [\varphi_i, \varphi_j](f) \pmod{f}$, we obtain:

Proposition. Let $[\vartheta] \in \ker\{H^1(\Theta_U) \to H^1(\Theta_e|U)\}$ with $\vartheta_{ij} = \varphi_j - \varphi_i$. The first obstruction to extend the infinitesimal deformation is the class $\frac{1}{2}[\varphi_i, \varphi_j] \in H^1(N_U)$.

Remark. To find the versal deformation of X, one could try to deform the U_i; as U_i is smooth, a deformation is given by a coordinate transformation $x + \varphi_i$. Considered as vector field, we get by Taylor's formula the equation

$$f(x + \varphi_i) = (\exp \varphi_i)(f) .$$

The transformation $x + \varphi_i$ will in general not be defined on U_i, for this one needs to shrink (cf. the constructions in [Pal, §5]). The (formal) power series $(\exp \varphi_i)(f)$ is defined on U_i. In the following we juggle with formulas without paying attention to such problems.

The condition that the deformations of the U_i fit together, is that there exist $\Phi_i = (\varphi_i, A_i)$ as above, such that for all i and j

$$(\exp \Phi_i)(f) = (\exp \Phi_j)(f) ,$$

or

$$\log(\exp(-\Phi_i)\exp(\Phi_j))(f) = 0 .$$

By the Baker-Campbell-Hausdorff formula $\log(\exp(-\Phi_i)\exp(\Phi_j)) + \Phi_i - \Phi_j$ is expressible by commutators. We want to translate the condition into one on the φ_i.

To compute $H^*(N_U)$ we use the mapping cone of $C^*(\mathcal{U}, \Theta_I) \to C^*(\mathcal{U}, \Theta_e)$; in particular, cocycles are pairs $(\varphi_i, \vartheta_{jk})$ with $\vartheta_{ij} = \varphi_j - \varphi_i$. A pair $([\varphi], [\vartheta])$ defines a deformation of X if and only if $\vartheta_{ij} = \log(\exp(-\varphi_i)\exp\varphi_j)$. By polarisation we obtain a product

$$([\varphi] \smile [\psi])_{ij} = \log(\exp(-\varphi_i)\exp\psi_j) + \log(\exp(-\psi_i)\exp\varphi_j) .$$

With this product our deformation equation is

$$[\vartheta] = \tfrac{1}{2}[\varphi] \smile [\varphi] .$$

One can try to solve this equation with formal power series $\vartheta = \sum \vartheta^{(k)}t^k$ and $\varphi = \sum \varphi^{(k)}t^k$. This gives as second obstruction

$$\tfrac{1}{2}\left[\varphi_i^{(1)}, \varphi_j^{(2)}\right] + \tfrac{1}{2}\left[\varphi_i^{(2)}, \varphi_j^{(1)}\right] + \tfrac{1}{12}\left[\left[\varphi_i^{(1)}, \varphi_j^{(1)}\right], \varphi_i^{(1)} + \varphi_j^{(1)}\right] .$$

I originally obtained this formula from a direct obstruction calculation.

The interesting aspect of these formulas is that they do not involve explicit knowledge of the equations, defining X. It suffices to have generators of the local ring as functions on the ambient space. The disadvantage is that the computations are rather involved, already for first obstruction in the standard example, the cone over the rational normal curve of degree four. In that example it is not clear from the computations, that the higher order obstructions vanish.

11 The projection method

In a series of papers Theo DE JONG and Duco VAN STRATEN studied deformations of normal surface singularities by means of a generic projection to $(\mathbb{C}^3, 0)$. The image is a non-isolated hypersurface singularity with double curves. Classes of such singularities had been studied by SIERSMA and PELLIKAAN. Their deformation theory becomes finite dimensional if one requires that the transverse singularity is not deformed. The normalisation of the deformed singularity will then be a deformation of the original singularity. An application is the computation of the base space for rational quadruple points [JS4].

To give an infinitesimal description it is better to avoid the normalisation and use an equivalent condition. The same ideas can then also be applied to study curve singularities.

Let $\widetilde{X}$ be a normal isolated surface singularity, which projects onto a hypersurface germ X with reduced, codimension one singular locus Σ. The curve Σ can be defined by the conductor ideal $\mathfrak{c} = \mathcal{H}om_X(\mathcal{O}_{\widetilde{X}}, \mathcal{O}_X)$ in $\mathcal{O}_X$. This allows one to reconstruct $\widetilde{X}$, because $\mathcal{O}_{\widetilde{X}} = \mathcal{H}om_X(\mathfrak{c}, \mathcal{O}_X)$. The fact that $\mathcal{O}_{\widetilde{X}}$ has a ring structure, is equivalent to the *Ring Condition*

$$\text{(R.C.)}: \qquad \mathcal{H}om_X(\mathfrak{c}, \mathfrak{c}) \overset{\approx}{\longrightarrow} \mathcal{H}om_X(\mathfrak{c}, \mathcal{O}_X).$$

This condition makes sense over any basis S, so we can set up a deformation theory along the same lines as deformations of singularities.

Various statements about the relation between such deformations and deformations of the original singularity hold in greater generality. Let $X \to Y$ be a finite, surjective and generically injective map. A deformation of this map is a map $X_S \to Y_S$. Let $\text{Def}(X \to Y)$ be the corresponding deformation functor. Let Σ be the subspace of Y defined by the conductor ideal $\mathfrak{c} = \mathcal{H}om_Y(\mathcal{O}_X, \mathcal{O}_Y)$. Then $\mathcal{O}_X = \mathcal{H}om_Y(\mathfrak{c}, \mathcal{O}_Y)$. The condition (R.C.) is now

$$\mathcal{H}om_Y(\mathfrak{c}, \mathfrak{c}) \overset{\approx}{\longrightarrow} \mathcal{H}om_Y(\mathfrak{c}, \mathcal{O}_Y).$$

Define the functor $\text{Def}(\Sigma \hookrightarrow Y, \text{R.C.})$ as functor of deformations for which the ideal of Σ_S in Y_S satisfies the condition (R.C.). Under suitable conditions, which are made explicit in the following two theorems, this functor is closely related to the deformation functor of X.

Theorem [JS3, (1.1)]. *Let $X \to Y$ be a finite, surjective and generically injective map. Let Σ be the subspace of Y defined by the conductor ideal $\mathfrak{c} = \mathcal{H}om_Y(\mathcal{O}_X, \mathcal{O}_Y)$. Assume that X is Cohen-Macaulay and Y is Gorenstein. Then there is a natural equivalence of functors $\mathrm{Def}(X \to Y) \longrightarrow \mathrm{Def}(\Sigma \hookrightarrow Y, \mathrm{R.C.})$.*

Theorem [JS3, (1.16)]. *Let $X \to Y$ be a generically injective map. Assume that X is Cohen-Macaulay and Y is a hypersurface. Then the natural transformation of functors $\mathrm{Def}(X \to Y) \longrightarrow \mathrm{Def}(X)$ is smooth.*

If X is a surface singularity, and $X \to Y$ a generic linear projection to a hypersurface in $(\mathbb{C}^3, 0)$, then Σ is a reduced curve singularity. In this case there is an alternative, more geometric description of the (R.C.) deformations of Y.

Consider *diagrams* over a base space S:

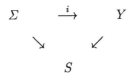

Normally we say that $\Sigma \to Y$ is a diagram over S, thereby suppressing the map i and in the notation also the base space. A *deformation* of a diagram $\Sigma_0 \to Y_0$ (over the point 0) is a diagram $\Sigma \to Y$ over $(S, 0)$, such that Σ and Y are flat over S, and $\Sigma_0 \to Y_0$ is isomorphic to the fibre over 0. The deformations of diagrams form a fibred groupoid, and lead to a homogeneous deformation functor $\mathrm{Def}(\Sigma_0 \to Y_0)$, cf. Chap. 6.

Definition [Rim1, Sect. 5]. Let $Y \to S$ be a flat mapping of relative dimension n. The critical space $\mathcal{C} := \mathcal{C}_{Y/S}$ is the space, defined by the n-th Fitting ideal $F_n(\Omega^1_{Y/S})$ of the sheaf of relative differentials.

Definition [JS2]. A diagram $\Sigma \to Y$ over S is *admissible*, if the map $i \colon \Sigma \to Y$ factorises over the inclusion map $\mathcal{C}_{Y/S} \hookrightarrow Y$. A deformation $\Sigma \to Y$ of a diagram $\Sigma_0 \to Y_0$ is admissible, if $\Sigma \to Y$ is an admissible diagram.

The admissible deformations form again a fibred groupoid, and lead to the deformation functor $\mathrm{Def}(\Sigma_0, Y_0)$.

Lemma (cf. [JS3, (1.7)]). *The functor $\mathrm{Def}(\Sigma_0, Y_0)$ is a homogeneous subfunctor of the functor $\mathrm{Def}(\Sigma_0 \to Y_0)$. If $\mathrm{Def}(\Sigma_0, Y_0)(\mathbb{C}[\varepsilon])$ is finite-dimensional, then a formal semi-universal deformation exists.*

In [JS2] conditions of various degrees of generality are given, under which the forgetful transformation $\mathrm{Def}(\Sigma_0, Y_0) \to \mathrm{Def}(Y_0)$ is injective. This holds for example, if Σ_0 and Y_0 are both Cohen-Macaulay, and furthermore $\dim \mathrm{Supp}\left(I_0 / F_n(\Omega^1_{Y_0})\right) < \dim \Sigma_0$, where I_0 is the ideal of Σ_0. In particular, if Y_0 is a hypersurface germ, defined by a function $f \in \mathcal{O}_{n+1}$, then

$F_n(\Omega^1_{Y_0}) = (f, J(f))$, with $J(f)$ the Jacobian ideal $(\partial_i f)$, and the condition becomes that Σ_0 is Cohen-Macaulay of positive dimension, and that $\dim_{\mathbb{C}} (I_0/(f, J(f)))$ is finite. If I_0 is reduced, then Y_0 has generic transverse A_1-singularities, and the admissibility condition ensures that this remains so under deformation.

Theorem [JS3, (1.2) and (1.3)]. *Let $\Sigma_0 \to Y_0$ be an admissible diagram, with Y_0 a hypersurface and Σ_0 Cohen-Macaulay of codimension two. Suppose that furthermore Σ_0 is reduced. Let $\Sigma \to Y$ be any deformation over a base $(S, 0)$ of this diagram. Then the ideal I of Σ satisfies the condition (R.C.), if and only if the diagram $\Sigma \to Y$ is admissible. Therefore there is a natural equivalence of functors $\mathrm{Def}(\Sigma_0 \hookrightarrow Y_0, \mathrm{R.C.}) \to \mathrm{Def}(\Sigma_0, Y_0)$.*

We now describe the infinitesimal deformations and obstructions in this theory, following the down to earth approach in the original paper [JS1]. The later account [JS2] gives an improved but more abstract version. We restrict ourselves to the hypersurface case: let Y be defined by the function $f \in \mathcal{O}_{n+1}$. Let Σ be a subspace of Y defined by the ideal I. Later on we shall make the extra assumption that Σ is Cohen-Macaulay of codimension two. In that case the ideal I is generated by the maximal minors $\Delta_1, \ldots, \Delta_k$ of an $k \times (k-1)$ matrix M. Following DE JONG and VAN STRATEN, we write also in the general case the generators of I as row vector Δ.

We recall the definition of the *primitive ideal* of I [Pel1, Def. 1.1]:

$$\int I := \{\, g \in \mathcal{O}_{n+1} \mid (g, \partial_i g) \subset I \,\} .$$

The admissibility condition now becomes

$$\Sigma \subset \mathcal{C}_Y \iff f \in \int I .$$

We translate the last condition in explicit equations: there exist a column vector $\alpha \in \mathcal{O}^k$, and a vector ω of 1-forms, such that $f = \Delta\alpha$ and $df = \Delta\omega$. Similar equations have to hold in the deformed situation.

Let $N_\Sigma := \mathrm{Hom}_\Sigma(I/I^2, \mathcal{O}_\Sigma)$ be the normal module of Σ. We write an element $n \colon \Delta_i \mapsto n_i$ of the normal bundle as row vector n. Admissible first order infinitesimal deformations are given by the space of *admissible pairs*, which is a subset of $N_\Sigma \times N_Y$:

$$\mathcal{A} = \left\{ (n, g) \in N_\Sigma \times N_Y \; \middle| \; \begin{array}{c} g = n\alpha + \Delta\alpha' \text{ and } dg = n\omega + \Delta\omega' \\ \text{for some } \alpha' \text{ and } \omega' \end{array} \right\} .$$

The equations are a direct consequence of the formula

$$f + \varepsilon g = (\Delta + \varepsilon n)(\alpha + \varepsilon\alpha')$$

and a similar one for $d(f + \varepsilon g)$. They lead us to consider three maps. First of all, we have the *evaluation map* $ev_f \colon N \to \mathcal{O}_\Sigma$, given by $n \mapsto n(f)$. In terms of α we can write $ev_f(n) = n\alpha$; the abstract definition shows that this representation is independent of the specific choice of α, which also easily can be seen directly. Secondly, the formula $n \mapsto n\omega$ defines a map $ev_{df}: N \to \Omega^1_{n+1} \otimes \mathcal{O}_\Sigma$. Finally, there is a map $w_f: N \to \Omega^1_\Sigma$, given by $w_f(n) = d(n\alpha) - n\omega$.

Lemma [JS1, (3.5)]. *There are exact sequences*

$$0 \longrightarrow P_\Sigma(\mathcal{A}) \longrightarrow N_\Sigma \xrightarrow{w_f} \Omega^1_\Sigma ,$$

$$0 \longrightarrow \int I/(f) \longrightarrow \mathcal{A} \longrightarrow P_\Sigma(\mathcal{A}) \longrightarrow 0 ,$$

where P_Σ is the projection on the first factor.

To obtain the tangent space to our deformation functor $\mathrm{Def}(\Sigma, Y)$, we divide out by the infinitesimal automorphisms on the ambient space:

$$T^1(\Sigma, Y) = \mathcal{A}/\{ (\vartheta(\Delta), \vartheta(f)) \mid \vartheta \in \Theta_{n+1} \} .$$

By [JS1, (3.6)] the maps ev_f and ev_{df} descend to maps $ev_f: T^1_\Sigma \to \mathcal{O}_\Sigma$ and $ev_{df}: T^1_\Sigma \to \Omega^1_\Sigma$. With the notations $J_\Sigma(f) = \{\theta(f) \mid \theta(I) \subset I\}$ and $T^1_\Sigma(Y) := \mathrm{Im}(T^1(\Sigma, Y) \to T^1_\Sigma)$, we get exact sequences

$$0 \longrightarrow T^1_\Sigma(Y) \longrightarrow T^1_\Sigma \xrightarrow{w_f} \Omega^1_\Sigma ,$$

$$0 \longrightarrow \int I/(f, J_\Sigma(f)) \longrightarrow T^1(\Sigma, Y) \longrightarrow T^1_\Sigma(Y) \longrightarrow 0 .$$

Obstructions come in two types. Firstly, there are the obstructions to extend deformations of Σ. They vanish in the main application, because $T^2_\Sigma = 0$ for a positive dimensional Cohen-Macaulay codimension two singularity Σ; in the zero-dimensional case Σ is at least unobstructed. Secondly, we can encounter obstructions, while trying to lift the equations $f = \Delta a$ and $df = \Delta\omega$; for details, see [JS1, (3.B)]. The relevant vector space is $T^2(\Sigma, Y) := \mathrm{coker}(w: T^1_\Sigma \to \Omega^1_\Sigma)$. If Σ is reduced, then the obstruction lands in fact in $\mathrm{Tors}(\Omega^1_\Sigma)/w(T^1_\Sigma)$, which is finite dimensional if Σ has an isolated singularity.

Example [JS2, (3.15)]. The following example is due to PELLIKAAN [Pel2, 2.4]. It was the starting point for DE JONG and VAN STRATEN. Let

$$f = (yz)^2 + (zx)^2 + (xy)^2 .$$

The singular locus is given by $\Delta = (yz, xz, xy)$. We can write $f = \Delta\Delta^t$ and $df = 2\Delta d\Delta^t$ so $\alpha = \Delta^t$ and $\omega = 2d\Delta^t$. The map w_f is the zero map: $w_f(n) = (dn)\Delta^t - n(d\Delta)^t = 0 \in \Omega^1_\Sigma$, so $T^1_\Sigma(Y) = T^1_\Sigma$. For the versal deformation of Σ we can take the equations

$$\Delta_S = (yz + t_1 y + t_1 t_3, xz + t_2 z + t_1 t_2, xy + t_3 x + t_2 t_3) .$$

One has $J_\Sigma(f) = (x\frac{\partial}{\partial x}f, y\frac{\partial}{\partial y}f, z\frac{\partial}{\partial z}f, yz\frac{\partial}{\partial x}f, xz\frac{\partial}{\partial y}f, xy\frac{\partial}{\partial z}f)$, so

$$(f, J_\Sigma(f)) = ((yz)^2, (zx)^2, (xy)^2, xyz^3, xy^3z, x^3yz) .$$

As $\int I = (xyz) + I^2$ we obtain $\{3xyz, 2x^2yz, 2xy^2z, 2xyz^2\}$ as basis for $\int I/(f, J_\Sigma(f))$. The infinitesimal deformations are therefore given by the following formulas, taken modulo $\mathfrak{m}_S^2$, where $\mathfrak{m}_S = (t_0, t_1, t_2, t_3, s_1, s_2, s_3)$ is the maximal ideal of the parameter space. Consider the matrix

$$H = \begin{pmatrix} 1 & s_2 & s_3 \\ s_2 & 1 & s_1 \\ s_3 & s_1 & 1 \end{pmatrix} .$$

Then modulo $\mathfrak{m}_S^2$

$$\alpha_S = H\Delta_S^t + t_0(x, y, z)^t$$
$$f_S = \Delta_S \alpha_S$$
$$\omega_S = 2Hd\Delta_S^t + 3t_0(dx, dy, dz)^t$$
$$df_S = \Delta_S \omega_S .$$

We have already used the correct lift of Δ to second order. We now compute modulo terms in $\mathfrak{m}_S^3$ that

$$df_S - \Delta_S \omega_S = t_0 t_1 (xdy - 2ydx) + t_0 t_2 (ydz - 2zdy) + t_0 t_3 (zdx - 2xdz) .$$

This is an element of $\Omega_\Sigma^1 \otimes (\mathfrak{m}_S^2/\mathfrak{m}_S^3)$. The three 1-forms $xdy - 2ydx$, $ydz - 2zdy$ and $zdx - 2xdz$ form a basis of $\mathrm{Tors}(\Omega_\Sigma^1)$. We obtain therefore the base equations

$$t_0 t_1 = t_0 t_2 = t_0 t_3 = 0 .$$

They define a base space S with two components, over which the formulas above define a deformation, which is therefore the versal admissible deformation.

   It is no coincidence that the equations for the base space are the same as those for the versal deformation of the cone over the rational normal curve of degree four. In fact, our original hypersurface is a projection of that cone. We obtain the normalisation by setting $yz = x\xi$, $xz = y\eta$ and $xy = z\zeta$. The following equations hold:

$$\mathrm{Rank} \begin{pmatrix} \xi & z & y \\ z & \eta & x \\ y & x & \zeta \end{pmatrix} \leq 1$$

and the equation f can be written as $xyz(\xi + \eta + \zeta)$, so we have the hyperplane section $\xi + \eta + \zeta = 0$ of the Veronese cone. The t_0-deformation corresponds to $\xi + \eta + \zeta + t_0$. The total space of this deformation is the Veronese cone and its projection is the cone over the Steiner Roman surface. A general fibre is then a Roman surface in affine 3-space. Figure 11.1 shows the real points, which include the singular locus consisting of the coordinate axes. The (t_1, t_2, t_3)-deformations correspond to the Artin component. As our equations are not the usual ones for the cone over the rational normal curve we get

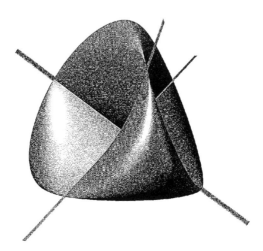

Fig. 11.1. The Steiner Roman surface

rather complicated equations. We normalise by setting $yz + t_1y + t_1t_3 = x\xi - t_2\eta$ and its cyclic permutations. Then $f_s = (xyz - t_1t_2t_3)(\xi + \eta + \zeta)$. Finally we have the s_i-deformations which change the hyperplane section to $\xi + \eta + \zeta + 2s_1x + 2s_2y + 2s_3z$. This is a trivial deformation of the cone; alternatively it can be obtained by changing the projection direction.

As the example shows it is easy to extend infinitesimal deformations given by perturbations lying in I^2. This observation leads to the the notion of I^2-equivalence for an ideal $I \subset \mathcal{O}_{n+1}$ [JS2, Def. 1.14]: two functions $f^{(1)}$ and $f^{(2)}$ are I^2-equivalent, if and only if $f^{(1)} - f^{(2)} \in I^2$, and two hypersurfaces are I^2-equivalent, if they can be defined by I^2-equivalent functions. This leads to I^2-equivalence of deformations, and the functor of admissible deformations modulo I^2, defined by $M(\Sigma, Y)(S) = \mathrm{Def}(\Sigma, Y)(S)/ \sim$. By [JS2, 1.16] the natural forgetful transformation $M(\Sigma, Y) \to \mathrm{Def}(\Sigma, Y)$ is smooth. In [JS4] this is used to prove that the base space of a rational quadruple point is up to a smooth factor equal to a certain space $B(n)$ where n is an invariant of the singularity, the number of virtual quadruple points: the maximal number of quadruple points into which the singularity deforms (see also p. 101 and p. 119).

We want to apply the projection method for curve singularities. So we are no longer interested in the normalisation, but we have still a finite generically injective map $X \to Y$. The relevant deformation theory is described in [JS2] for the case Σ reduced, where it is used to obtain a smaller obstruction space. The key observation is that for reduced Σ the evaluation map $ev_f : T^1_\Sigma \to \mathcal{O}_\Sigma$

is the zero map [JS2, (3.1)]. This condition can also be formulated in the relative case: let $\Sigma_S \to Y_S$ be a deformation of $\Sigma \to Y$ over S, given by $F = \Delta_S \alpha_S$. Then $ev_F \colon N_S \to \mathcal{O}_{\Sigma_S}$ is given by $n_S \mapsto n_S \alpha_S$.

Proposition [JS2, (3.4)]. *Suppose that Σ is reduced. The diagram $\Sigma_S \to Y_S$ is admissible if and only if the map ev_F is the zero map.*

For non-reduced Σ admissibility is no longer equivalent with ev_F being the zero map, but it is the latter property which is equivalent with the condition (R.C.), as is proved in [JS3, (1.12)]. Altogether we obtain

Theorem. *Let $X \to Y$ be a finite, generically injective map with $Y = \{f = 0\}$ a hypersurface and let Σ be defined by the conductor ideal. Suppose Σ is Cohen-Macaulay of codimension two. Let $\Sigma_S \to Y_S$ be any deformation over a base $(S, 0)$, with Y_S defined by a function F. The ideal I_S of Σ_S satisfies the condition (R.C.), if and only if the map ev_F is the zero map.*

This leads us to consider the functor $\mathrm{Def}(\Sigma \hookrightarrow Y, ev)$ of deformations where the map ev_F is the zero map. At this stage we do not need to assume that Σ is Cohen-Macaulay of codimension two, but we do assume that the deformations of Σ are unobstructed. The map ev_F is zero, if for every normal vector n_S there exists a vector γ_S on the ambient space, satisfying

$$n_S \alpha_S + \Delta_S \gamma_S = 0 . \tag{11.1}$$

This is the basic equation, which leads to an obstruction theory, quite similar to the one above.

Definition. Let $I \subset \mathcal{O}_{n+1}$ be an ideal. The *evaluation ideal* of I is the ideal:

$$I^{(ev)} = \{ g \in I \mid ev_g = 0 \} .$$

We remark that $I^2 \subset I^{(ev)} \subset \int I$.

First order infinitesimal deformations are given by the space of (ev)-*admissible pairs*

$$\mathcal{A}_{ev} = \left\{ (n, g) \in N_\Sigma \times N_Y \;\middle|\; \begin{array}{l} f + \varepsilon g = (\Delta + \varepsilon n)(\alpha + \varepsilon \alpha') \text{ for some } \alpha' \\ ev_{(f + \varepsilon g)} = 0 \end{array} \right\} .$$

Because for a small surjection $S' \to S$ with kernel V (so $\mathfrak{m}_{S'} V = 0$) and a deformation $\Sigma_{S'}$ of Σ, one has an exact sequence

$$0 \longrightarrow V \otimes N_{\Sigma_{S'}} \longrightarrow N_{\Sigma_{S'}} \longrightarrow N_{\Sigma_S} \longrightarrow 0$$

with $V \otimes N_{\Sigma_{S'}} \cong V \otimes_{\mathbb{C}} N_\Sigma$ [JS2, Prop. 3.5], we can write the condition $ev_{(f + \varepsilon g)} = 0$ explicitly as $(m + \varepsilon m_1)(\alpha + \varepsilon \alpha_1) + (\Delta + \varepsilon n)(\gamma + \varepsilon \gamma_1) = 0$ for every normal vector $(m + \varepsilon m_1)$.

We recall the definition of the *Hessian form* $H \colon N_\Sigma \otimes N_\Sigma \to \mathcal{O}_\Sigma$ [JS2, 3.6]. For every $n \in N_\Sigma$ and every relation r (i.e. $\Delta r = 0$) we find a $s(n)$

such that $nr + \Delta s(n) = 0$. Because $ev_f = 0$, there is also a $\gamma(n)$ such that $n\alpha + \Delta\gamma(n) = 0$. As Σ is unobstructed, given $m, n \in N_\Sigma$, there exist a p and t with

$$pr + ms(n) + ns(m) + \Delta t = 0 .$$

Now put $\boldsymbol{H}(n, m) = p\alpha + m\gamma(n) + n\gamma(m)$. The form $\boldsymbol{H}$ induces a map $\boldsymbol{h}: N_\Sigma \to N_\Sigma^*$.

Just as in [JS1, Cor. 3.35], we have $\ker(\boldsymbol{h}: N_\Sigma \to N_\Sigma^*/I) = P_\Sigma(\mathcal{A}_{ev})$, and an exact sequence

$$0 \longrightarrow I^{(ev)}/(f) \longrightarrow \mathcal{A}_{ev} \longrightarrow P_\Sigma(\mathcal{A}_{ev}) \longrightarrow 0 .$$

To obtain the tangent space $T^1(\Sigma \hookrightarrow Y, ev)$ to our deformation functor $\mathrm{Def}(\Sigma \hookrightarrow Y, ev)$, we again divide out by the infinitesimal automorphisms. Similar to [JS2, Propositions 3.10 and 3.11], one has:

Proposition. *Let the assumptions be the same as before. There is an exact sequence*

$$0 \to I^{(ev)}/(f, J_\Sigma(f)) \longrightarrow T^1(\Sigma \hookrightarrow Y, ev) \longrightarrow$$
$$\longrightarrow T_\Sigma^1 \xrightarrow{\ \boldsymbol{h}\ } N^*/I \longrightarrow T^2(\Sigma \hookrightarrow Y, ev) \to 0 .$$

The obstructions to extending (ev)-admissible deformations lie in the space $T^2(\Sigma \hookrightarrow Y, ev)$.

Theorem. *If Σ has an isolated singularity, the functor $\mathrm{Def}(\Sigma \hookrightarrow Y, ev)$ admits an (analytic) semi-universal deformation.*

Sketch of proof. As we should have remarked before, the functor satisfies the Schlessinger condition (S1'), as an explicit computation with equation (11.1), similar to the one for the ordinary deformation functor, shows. The condition on the singularities of Σ implies (S2). The proof is again completed by applying Grauert approximation (see Chap. 9). □

Also in this situation we can consider I^2-equivalence. We get the functor of (ev)-admissible deformations modulo I^2, defined by $M(\Sigma \hookrightarrow Y, ev)(S) = \mathrm{Def}(\Sigma \hookrightarrow Y, ev)(S)/\sim$.

Proposition. *The natural forgetful transformation*

$$\mathrm{Def}(\Sigma \hookrightarrow Y, ev) \to M(\Sigma \hookrightarrow Y, ev)$$

is smooth.

Proof. We argue as in the proof of [JS2, 1.16]. Smoothness of a transformation of functors $F \to G$ means that for all surjections $T \to S$ the natural map $F(T) \to F(T) \times_{G(T)} G(S)$ is also surjective. So let $f_S = \Delta_S \alpha_S$ and $g_S = \Delta_S \beta_S$ be I^2-equivalent. Then $\alpha_S - \beta_S = A_S \Delta_S^t$ for some matrix A_S. Let $T \to S$ be a surjection and $n_T \alpha_T + \Delta_T \gamma_T$ be a lift of $n_S \alpha_S + \Delta_S \gamma_S$. Choose any lift A_T of A_S. Then $n_T(\alpha_T - A_T \Delta_T^t) + \Delta_T(\gamma_T + A_T^t n_T^t)$ is a lift of $n_S \beta_S + \Delta_S(\gamma_S + A_S^t n_S^t)$. □

Now we return to the case of a hypersurface Y, obtained as projection of a singularity X, such that Σ is Cohen-Macaulay of codimension two. Then the functor of (R.C.)-deformations is the same as that of (ev)-admissible deformations. The result of a computation is a base space S and a matrix $\widetilde{M}_S := (M_S, \alpha_S)$. Out of these data equations for a deformation of X can be constructed [Cat, MP]. We describe this in the absolute case.

The ideal of Y in the regular local ring $\mathcal{O}_{n+1}$ is generated by $f = \det(\widetilde{M})$, as we have seen. But the matrix $\widetilde{M}$ does more: it defines a presentation

$$0 \longrightarrow \mathcal{O}_{n+1}^k \xrightarrow{\ \widetilde{M}\ } \mathcal{O}_{n+1}^k \longrightarrow \mathcal{O}_X \longrightarrow 0$$

of $\mathcal{O}_X$ as $\mathcal{O}_{n+1}$-module; generators of $\mathcal{O}_X$ are $g_1, \ldots, g_{k-1}, g_k = 1$. Let $\widetilde{M}^* = \bigwedge^{k-1} \widetilde{M}$ be the Cramer matrix of $\widetilde{M}$. The generators g_i of $\mathcal{O}_X = \mathrm{Hom}_Y(I, I)$ act on I as multiplication with $\widetilde{M}_{i1}^*/\widetilde{M}_{k1}^*$ [MP, Prop. 3.14]. One has relations $g_i g_j = \sum h_{ij}^r g_r$ with $h_{ij}^r \in \mathcal{O}_Y$. Then X can be embedded in $\mathbb{C}^{n+1} \times \mathbb{C}^{k-1}$ with equations [MP, 3.11]

$$\sum_{j=1}^{k-1} \widetilde{M}_{ij} v_j + \widetilde{M}_{ik} = 0 , \qquad 1 \le i \le k ,$$

$$v_i v_j = \sum_{r=1}^{k-1} h_{ij}^r v_r + h_{ij}^k , \qquad 1 \le i, j \le k - 1 .$$

We apply the preceding theory to curve singularities. Let $X \subset \mathbb{C}^N$ be a curve and $\pi \colon X \to Y$ a degree one map to a plane curve singularity Y, e.g., a (general) projection. In this situation the conductor ideal $\mathfrak{c} = \mathrm{Ann}_{\mathcal{O}_Y}(\mathcal{O}_X/\mathcal{O}_Y)$ defines a fat point Σ in the plane, which is indeed Cohen-Macaulay of codimension 2.

Lemma [MP, Thm. 3.6]. *The multiplicity of Σ is equal to $\delta(Y) - \delta(X)$.*

Proof. This Lemma is a generalisation of the equality $n = 2\delta$ [Ser, IV.11]. Because $\mathcal{O}_X/\mathcal{O}_Y$ is dual to ω_Y/ω_X, both modules have the same annihilator, so a function f satisfies $f \in \mathfrak{c}$ if and only if $f\alpha \in \omega_X$ for all $\alpha \in \omega_Y$. As Y is a plane curve, ω_Y is a free module with generator α. The map $\mathcal{O}_X/\mathfrak{c} \to \omega_Y/\omega_X$, defined by $f \mapsto f\alpha$, is an isomorphism. $\qquad\square$

We denote by δ_{pr} the delta-invariant of a general plane projection of a curve singularity. One has $k - 1 \ge N - 2$: the equations described above give an embedding in $\mathbb{C}^{k+1}$.

Example. We treat in detail the singularity $X = L_4^4$. We take for X the following determinantal equations

$$\begin{vmatrix} x & y & z & t \\ y & z & t & -x \end{vmatrix} .$$

The projection Y on the (x,y)-plane is given by the function $f = x^4 + y^4$; the subspace Σ has multiplicity 3 (because $\delta_{pr} - \delta = 3$) and $k - 1 \geq 2$, so the conductor ideal is $I = (x^2, xy, y^2)$. The versal deformation of Σ is given by the maximal minors Δ_i of

$$\begin{vmatrix} y & e_{01} \\ -(x + e_{10}) & y + e_{11} \\ -e_{20} & -x \end{vmatrix}.$$

One has $I^{(ev)} = (x,y)^3 = \int I$, and $J_\Sigma(f) = (x^4, x^3 y, xy^3, y^4)$, so we can take the infinitesimal deformation $x^4 + y^4 + \lambda x^2 y^2 + d_0 x^3 + d_1 x^2 y + d_2 xy^2 + d_3 y^3$ of f. We will take $\lambda = 0$, so we consider only the negative part of the (ev)-admissible versal deformation of Y: because λ gives a deformation in I^2, we know how to adapt the formulas to include it. Furthermore T_X^1 has no part of degree 0, so the λ deformation leads to a trivial deformation of X.

The normal module N of Σ is generated by 6 elements. We write them as (6×3) matrix m. An element of N^* is then represented by a column vector, whose entries are subject to conditions, coming from the relations between the generators of N; in this case the only relations are that every element of N is annihilated by the maximal ideal, so an element of N^* is a column vector with entries in $\mathfrak{m}$. The 3 columns of the matrix represent the embedding of I in N^*.

Let S be a smooth 8-dimensional germ, with coordinates (e, d), as given above. We can find a vector α_S and a matrix γ_S such that $m_S \alpha_S - \gamma_S^t \Delta_S^t \equiv 0$ (mod $(e, d)^2$). This shows that the map $T^1(\Sigma \hookrightarrow Y, ev) \to T_\Sigma^1$ is surjective, so $h = 0$. To determine the obstruction we look at the quadratic terms in $m_S \alpha_S + \gamma_S^t \Delta_S^t$, i.e. we read our formulas modulo third powers of the deformation parameters:

$$\begin{pmatrix} 0 & x & y + e_{11} \\ 0 & -e_{20} & x + e_{10} \\ -x & 0 & -e_{01} \\ e_{20} & 0 & y \\ -y - e_{11} & e_{01} & 0 \\ -x - e_{10} & -y & 0 \end{pmatrix} \begin{pmatrix} x^2 + d_0 x \\ d_1 x + d_2 y \\ y^2 + d_3 y \end{pmatrix}$$

$$- \begin{pmatrix} d_1 & d_2 - e_{01} & y + d_3 \\ 0 & y + d_3 & e_{10} \\ -x + e_{10} - d_0 & e_{20} & -e_{01} \\ e_{20} & -e_{01} & y - e_{11} + d_3 \\ -e_{11} & -x - d_0 & 0 \\ -x - d_0 & -d_1 + e_{20} & -d_2 \end{pmatrix} \begin{pmatrix} x^2 + e_{10} x + e_{20} y + e_{20} e_{11} \\ yx - e_{20} e_{01} \\ y^2 + e_{11} y + e_{01} x + e_{01} e_{10} \end{pmatrix}$$

$$\equiv \begin{pmatrix} (-d_1 e_{10} - d_3 e_{01}) x - (d_1 e_{20} + e_{10} e_{01}) y \\ -(d_1 e_{20} + e_{10} e_{01}) x + (-e_{20} d_2 + e_{10} d_3 - e_{10} e_{11} + e_{20} e_{01}) y \\ (e_{10} d_0 - e_{10}^2 + e_{01}^2 + e_{20} e_{11}) x + (-e_{01} d_3 + d_0 e_{20} - e_{10} e_{20} + e_{01} e_{11}) y \\ (d_0 e_{20} - e_{10} e_{20} + e_{11} e_{01} - d_3 e_{01}) x + (-e_{20}^2 + e_{11}^2 - e_{11} d_3 - e_{01} e_{10}) y \\ (-e_{11} d_0 + e_{01} d_1 + e_{11} e_{10} - e_{20} e_{01}) x + (e_{11} e_{20} + e_{01} d_2) y \\ (e_{01} d_2 + e_{20} e_{11}) x + (d_0 e_{20} + d_2 e_{11}) y \end{pmatrix}.$$

The resulting vector represents an element of $N^* \otimes (\mathcal{O}_S/\mathfrak{m}_S^3)$. We reduce it with I, i.e., with the columns of the undeformed matrix m; more precisely, we add $(e_{01}d_2 + e_{20}e_{11}, d_1e_{10} + d_3e_{01}, d_1e_{20} + e_{10}e_{01})^t$ to α_S. We obtain as obstruction vector

$$
\begin{pmatrix}
0 \\
(-e_{20}d_2 + e_{10}d_3 - e_{10}e_{11} + e_{20}e_{01})y \\
(e_{10}d_0 - e_{10}^2 + e_{01}^2 - e_{01}d_2)x + (-e_{01}d_3 + d_0e_{20} - e_{10}e_{20} + e_{01}e_{11})y \\
(d_0e_{20} - e_{10}e_{20} + e_{11}e_{01} - d_3e_{01})x + (-e_{20}^2 + e_{11}^2 - e_{11}d_3 + d_1e_{20})y \\
(-e_{11}d_0 + e_{01}d_1 + e_{11}e_{10} - e_{20}e_{01})x \\
(d_0e_{20} + d_2e_{11} - d_1e_{10} - d_3e_{01})y
\end{pmatrix}.
$$

The coefficients vanish, if the minors of the following matrix vanish. We recover the fact that the base space is the cone over the Segre embedding of $\mathbb{P}^1 \times \mathbb{P}^3$:

$$
\begin{vmatrix}
d_0 - e_{10} & d_1 - e_{20} & d_2 - e_{01} & d_3 - e_{11} \\
e_{01} & e_{11} & e_{10} & e_{20}
\end{vmatrix}.
$$

We can now compute equations for the deformation of $L_4^4 \subset \mathbb{C}^4$. They are not determinantal, but we can write each equation as the sum of two minors; in the following symbol one has to add corresponding minors:

$$
\begin{vmatrix}
x & y + d_3 - e_{11} & z + d_2 - e_{01} & t \\
y + e_{11} & z + e_{01} & t - d_1 & -x - d_0
\end{vmatrix}
+
\begin{vmatrix}
t + e_{20} & x + e_{10} & y + d_3 & z + d_2 \\
e_{01} & e_{20} & e_{10} & e_{11}
\end{vmatrix}.
$$

I also did a direct computation of the versal deformation of the four lines in determinantal form; the one presented here, using the projection, is much easier, especially if one is only interested in the base space (i.e. one does not determine the equations for the deformation of X). However, the direct computation for L_4^4 as coordinate axes in Chap. 3 is even easier, due to the extra symmetry, which is lost in the determinantal form.

We use slightly different equations, and new letters for the deformation parameters, to make the formulas more symmetric. The way of writing the equations as sum of two determinantal symbols suggests the following lift of the four relations, obtained by doubling the first row of the matrix defining the singularity:

$$
\begin{vmatrix}
x & y & z & t \\
\cdots & \cdots & \cdots & \cdots \\
x & y & z & t \\
y + B & z + C & t + D & x + A
\end{vmatrix}
+
\begin{vmatrix}
x & y & z & t \\
\cdots & \cdots & \cdots & \cdots \\
t + d & x + a & y + b & z + c \\
c & d & a & b
\end{vmatrix}
$$

$$
+
\begin{vmatrix}
c & d & a & b \\
\cdots & \cdots & \cdots & \cdots \\
t & x & y & z \\
x + A & y + B & z + C & t + D
\end{vmatrix}
+
\begin{vmatrix}
c & d & a & b \\
\cdots & \cdots & \cdots & \cdots \\
z + c & t + d & x + a & y + b \\
b & c & d & a
\end{vmatrix}
$$

$$
= \begin{vmatrix} c & d & a & b \\ \cdots\cdots\cdots\cdots\cdots\cdots\cdots\cdots \\ t & x & y & z \\ A-a & B-b & C-c & D-d \end{vmatrix} + \begin{vmatrix} c & d & a & b \\ \cdots\cdots\cdots\cdots\cdots\cdots\cdots \\ c & d & a & b \\ y+b & z+c & t+d & x+a \end{vmatrix}.
$$

The maximal minors of the first matrix vanish, and for the remaining five the equality already holds for the minors of the 2×4 matrices, obtained by deleting the row (c, d, a, b); this is basically the identity

$$
\begin{vmatrix} a & b & c & d \\ e & f & g & h \end{vmatrix} + \begin{vmatrix} b & c & d & a \\ h & e & f & g \end{vmatrix} + \begin{vmatrix} c & d & a & b \\ g & h & e & f \end{vmatrix} + \begin{vmatrix} d & a & b & c \\ f & g & h & e \end{vmatrix} = 0 .
$$

Therefore the right hand side of the equation gives as base space

$$
\begin{vmatrix} A-a & B-b & C-c & D-d \\ c & d & a & b \end{vmatrix} .
$$

For the other four relations I have not found such a lift.

Theorem. Let X be a curve singularity of multiplicity 4 and embedding dimension 4. Let X_{pr} be a general plane projection. If $\delta(X_{pr}) - \delta(X) = 3$, then the base space of the versal deformation of X is $C_{\mathbb{P}^1 \times \mathbb{P}^3} \times \mathbb{C}^{\tau-8}$, the cone over the Segre embedding times a smooth factor.

Proof. Because $\delta(X_{pr}) - \delta(X) = 3$, and $t(\Sigma) \geq 2$, the conductor ideal is again $I = (x^2, xy, y^2)$; since X_{pr} has multiplicity 4, it is I^2-equivalent to the singularity $x^4 + y^4$ from the previous example. So up to a smooth factor the base space is the same. $\square$

The theorem applies in particular to *partition curves* [BC] of multiplicity four. Such curves have been introduced by various authors under different names (cf. [St5]). The first occurrence is in SERRE's book as curve defined by a 'module' [Ser, IV.4].

Definition. A *partition curve* X is a curve singularity, for which the conductor ideal is equal to the maximal ideal: $\mathfrak{c}_X = \mathfrak{m}_X$.

The general hyperplane section of a rational surface singularity is a partition curve. For this reason these curves where classified in [BC]. Recall that a reducible curve $X = X_1 \cup X_2$ is called *decomposable*, if the Zariski tangent spaces of X_1 and X_2 have only one point in common, notation: $X = X_1 \vee X_2$. Let X_n be the monomial curve with semigroup $n, n+1, \ldots, 2n-1$; it can be given parametrically by $z_i = t^{i+n-1}$, $i = 1, \ldots, n$, or by determinantal equations, the (2×2)-minors of

$$
\begin{vmatrix} z_1 & z_2 & \cdots & z_{n-1} & z_n \\ z_2 & z_3 & \cdots & z_n & z_1^2 \end{vmatrix} .
$$

Let $n = n_1 + \cdots + n_r$ be a partition of n. The corresponding partition curve is the wedge $X_{n_1,\ldots,n_r} = X_{n_1} \vee \cdots \vee X_{n_r}$. Every curve singularity with $\delta = n-1$

is isomorphic to one these curves. The partition $(1, \ldots, 1)$ gives the curve L_n^n, with equations $z_i z_j = 0$, $1 \leq i < j \leq n$. For the general partition curve one has coordinates $z_i^{(j)}$, $i = 1, \ldots, n_j$, $j = 1, \ldots, r$, and equations $z_k^{(i)} z_l^{(j)} = 0$ for $i \neq j$ and all k and l, plus for each $n_i > 1$ the corresponding determinantal equations in the $z_k^{(i)}$.

A general partition curve of multiplicity n occurs in the versal deformation of the monomial curve X_n. Consider the deformation of X_n, given by

$$\begin{vmatrix} z_1 & z_2 & z_3 & \cdots & z_{n-1} & z_n \\ z_2 & z_3 + a_3 z_1 & z_4 + a_4 z_1 & \cdots & z_n + a_n z_1 & z_1(z_1 + a_1) \end{vmatrix} .$$

For all values of the parameters this still defines a curve of multiplicity n: if we project on the (z_1, z_2)-plane, we get the equation

$$z_1^{n+1} + a_1 z_1^n + a_n z_1^{n-1} z_2 + a_{n-1} z_1^{n-2} z_2^2 + \ldots + a_3 z_1^2 z_2^{n-2} - z_2^n = 0 .$$

The number of branches of this curve is given by the number of distinct roots of the initial part, the polynomial of degree n; the multiplicity of the root is the multiplicity of the corresponding branch of the partition curve. Our polynomial describes the versal deformation of $-z_2^n$, so all partition curves of multiplicity n appear in this family.

Example. We give the results of an explicit calculation for the monomial partition curve with projection $Y : x^5 + y^4 = 0$. The matrices Δ and m are the same as for L_4^4. We again consider only deformations of negative weight, so we take as first order deformation of Y the function $x^5 + y^4 + c_0 x^4 + c_1 x^3 y + c_2 x^2 y^2 + d_0 x^3 + d_1 x^2 y + d_2 x y^2 + d_3 y^3$. We find for α the matrix

$$\begin{pmatrix} x^3 + c_0 x^2 + d_0 x + d_2 e_{01} - c_1 e_{01} e_{20} + (c_0 - e_{10}) e_{11} e_{20} + e_{01} e_{20}^2 \\ c_1 x^2 + d_1 x + c_2 x y + d_2 y + d_3 e_{01} + d_1 e_{10} - c_1 e_{10}^2 - c_2 e_{01} e_{20} + c_1 e_{11} e_{20} \\ y^2 + d_3 y + e_{01} e_{10} + d_1 e_{20} - c_1 e_{10} e_{20} \end{pmatrix}$$

and as equations for the base space the minors of

$$\begin{vmatrix} d_0 - (c_0 - e_{10}) e_{10} - e_{20} e_{11} & d_1 - (c_0 - e_{10}) e_{20} - c_1 e_{10} & d_2 - e_{01} - c_1 e_{20} & d_3 - e_{11} \\ e_{01} & e_{11} & e_{10} & e_{20} \end{vmatrix} .$$

This space is by an obvious coordinate transformation isomorphic to $C_{\mathbb{P}^1 \times \mathbb{P}^3} \times \mathbb{C}^3$. This transformation simplifies the equations of the base space, but makes the vector α more complicated.

For a general plane projection of a partition curve of multiplicity 5 the conductor ideal is m^3. Every other curve of multiplicity 5 with the same conductor is I^2-equivalent with the projection of a universal curve of multiplicity 5. So we obtain:

Theorem. *Let X be a curve singularity of multiplicity 5 and embedding dimension 5. Let X_{pr} be a general plane projection. If $\delta(X_{pr}) - \delta(X) = 6$, then the base space of the versal deformation of X is up to a smooth factor isomorphic to the base of a partition curve of multiplicity 5. The partition is determined by the multiplicities of the roots of the degree 5 part of the equation of X_{pr}.*

Explicit equations for the base space of the monomial partition curve of multiplicity 5 are given in [St5].

12 Formats

Apart from the existence, hardly any general results are known about versal deformations of isolated singularities. Indeed, it seems that every imaginable pathology occurs: the base space $(S,0)$ is in general reducible, with components of varying dimension, including embedded components .... For special classes of singularities more can be said. Cohen-Macaulay singularities of codimension two are determinantal, as are all deformations: they are obtained by perturbing the entries of the defining $t \times (t+1)$ matrix; the number t can be characterised as number of generators of the dualising module. In particular, the base space is smooth. The Gorenstein case is included in this description, because for $t = 1$ we have a complete intersection. A Gorenstein codimension 3 singularity is Pfaffian [BE]: the ideal is generated by the Pfaffians (square roots of principal minors) of an antisymmetric $(2s+1) \times (2s+1)$ matrix, with $s = 1$ in the complete intersection case. Any minimal free resolution of the ideal has a $(2s + 1) \times (2s + 1)$ matrix in the middle, which can be made anti-symmetric by elementary matrix operations. Deformations are again obtained by deforming the matrix, and the base space is smooth.

In general the base space of the versal deformation is reducible. In a number of examples, most prominently the cone over the rational normal curve of degree four, the total space over each component separately can be given in a compact way by determinantal equations, or some variation of those. Following Miles REID we call this a *format*. In a naive interpretation a format is a way of writing or coding (efficiently) the equations of a singularity. The format is *flexible* [Re2], if an arbitrary deformation of the entries yields a flat deformation; this is the case if there is an automatic procedure to get the relations.

Example 1: determinantals. A singularity X is called *determinantal*, if its ideal is generated by the $t \times t$ minors of an $r \times s$ matrix, and X is of pure codimension $(r-t+1)(s-t+1)$. The condition on the codimension guarantees that X is Cohen-Macaulay, for it is the pull-back of the generic determinantal singularity, which is defined by the minors of a matrix of indeterminates (cf. [ACGH, II.4]).

Example 2: quasi-determinantals. This format is a generalisation of the determinantal format. Consider in the $2 \times k$ case the following symbol:

$$\begin{vmatrix} f_1 & f_2 & \cdots & f_{k-1} & f_k \\ g_1 & g_2 & \cdots & g_{k-1} & g_k \\ & h_{1,2} & \cdots & & h_{k-1,k} \end{vmatrix}$$

Its generalised minors are $f_i g_j - g_i (\prod_{\alpha=i}^{j-1} h_{\alpha,\alpha+1}) f_j$ [Rie3]. All two-dimensional quotient singularities can be written in this way [Rö, Kor. 4.2.2]; explicit equations were given by Constantin KAHN (unpublished). Ancus RÖHR characterised all rational surface singularities which can be written in quaside-terminantal format by the property that their resolution graph contains the graph of a cyclic quotient singularity of the same multiplicity as subgraph. Explicit equations were obtained by Theo DE JONG [Jo].

In [PR] a generalisation of the format of the maximal minors of the general $r \times s$ matrix is given. The construction is based on a degeneration of the generic determinantal, obtained from the *Hodge algebra structure*. Let R be a commutative ring and A a commutative R-algebra, and let $H \subset A$ be a finite partially ordered set, such that:

1. A is a free R-module with basis consisting *standard monomials* $x_1 \cdots x_s$, where $x_i \in H$ and $x_1 \leq \cdots \leq x_s$;
2. if $y_1, y_2 \in H$ are incomparable, and

$$y_1 y_2 = \sum r_i x_{i,1} \cdots x_{i,s_i} , \quad r_i \in R ,$$

is the expression of $y_1 y_2$ in the basis of standard monomials, then $x_{i,1} < y_1, y_2$ for all i.

Then A is a *Hodge algebra* (in the terminology of [CEP] an *ordinal* Hodge algebra) on H. If all the products $y_1 y_2$ are zero, then A is the *discrete* Hodge algebra on H, and there exists a degeneration of a general Hodge algebra A on H to the discrete Hodge algebra A_0, which can be constructed inductively in a finite number of steps, say n. In [PR] it is shown that there exists a Hodge algebra $\mathcal{A}$ on H over $R[t_1, \ldots, t_n]$, which gives for $t_i = 1$ the Hodge algebra A and for $t_i = 0$ the discrete Hodge algebra A_0. The total space of this deformation is then the more general format. In most applications the algebra A is rigid, so not only for $t_i = 1$, but for general values of the t_i the Hodge algebra is isomorphic to A.

Example 3: a degeneration of the cone over the Segre embedding of $\mathbb{P}^2 \times \mathbb{P}^2$. The equations of the Segre cone are the 2×2 minors of the generic 3×3 matrix M. They can also be obtained in a more complicated way [Re2, §5]: write $M = A + B$ as sum of a symmetric matrix A and a skew matrix B, and define

$$P = \begin{pmatrix} B & A \\ -A & -B \end{pmatrix}.$$

The 15 diagonal 4×4 Pfaffians of P are not linearly independent, but define the same ideal as the minors of M. We now introduce a new parameter t, and define

$$P(t) = \begin{pmatrix} B & A \\ -A & -tB \end{pmatrix}.$$

There are still only 9 linearly independent Pfaffians. As the Segre cone is rigid, we can write the new ideal in a neighbourhood of $t = 1$ as minors of a 3×3 matrix: put $u = \sqrt{t}$ (this may not be possible in an affine neighbourhood, but we work here locally in the complex topology), and let $M(u) = A + uB$. The proof that the ideal of the minors of $M(u)$ coincides with that of the Pfaffians of $P(u^2)$ uses in an essential way the fact that u is a unit; taking the limit $u \to 0$ in $M(u)$ gives only the six equations Rank $A \le 1$. Let

$$B = \begin{pmatrix} 0 & b_3 & -b_2 \\ -b_3 & 0 & b_1 \\ b_2 & -b_1 & 0 \end{pmatrix}$$

and write $b = (b_1, b_2, b_3)^t$. We get from $P(0)$ the equations

$$\text{Rank } A \le 1, \qquad Ab = 0.$$

This describes the image of the incidence variety in the join of $\mathbb{P}^2$ and its dual $\check{\mathbb{P}}^2$ under the map, which is the Veronese embedding on the first factor. The space is singular on the locus where the entries of A vanish, with as transverse singularity an A_1 (consider for instance the transverse slice $b = (1, 0, 0)^t$). The total space of the deformation $P(t)$ is rigid.

Example 4: rolling factors [Re2]. This format occurs often in connection with divisors on scrolls. We start with a k-dimensional rational normal scroll $S \in \mathbb{P} := \mathbb{P}^n$; the classical construction is to take k complementary linear subspaces L_i, spanning $\mathbb{P}$, with a parametrised rational normal curve $\phi_i : \mathbb{P}^1 \to C_i \subset L_i$ of degree $d_i = \dim L_i$ in it, and to take for each $p \in \mathbb{P}^1$ the span of the points $\phi_i(p)$. The degree of S is $d = \sum d_i = n - k + 1$. To give a coordinate description we take homogeneous coordinates $(s : t)$ on $\mathbb{P}^1$ and $(z^{(1)} : \cdots : z^{(k)})$ on the fibres. Coordinates on $\mathbb{P}$ are $z_j^{(i)} = z^{(i)} s^{d_i - j} t^j$, with $0 \le j \le d_i$, $1 \le i \le k$. We give the variable y_i the weight $-d_i$. The scroll S is then given by the minors of the matrix

$$\Phi = \begin{pmatrix} z_0^{(1)} & \cdots & z_{d_1-1}^{(1)} & \cdots & z_0^{(k)} & \cdots & z_{d_k-1}^{(k)} \\ z_1^{(1)} & \cdots & z_{d_1}^{(1)} & \cdots & z_1^{(k)} & \cdots & z_{d_k}^{(k)} \end{pmatrix}.$$

We now consider a divisor on S in the linear system $|aH - bR|$, where the hyperplane class H and the ruling R generate the Picard group of S. When we speak of degree on S this will be with respect to H. The divisor can be given by one bihomogeneous equation $P(s, t, z^{(i)})$ of degree a in the $z^{(i)}$, and total degree $-b$. By multiplying $P(s, t, z^{(i)})$ with a polynomial of degree b in $(s : t)$ we obtain an equation of degree 0, which can be expressed as polynomial of degree a in the $z_j^{(i)}$. This expression is not unique, but the difference of two expressions lies in the ideal of the scroll. By the obvious choice, multiplying with $s^{b-m}t^m$, we obtain $b+1$ equations P_m. In the transition from the equation P_m to P_{m+1} we have to increase by one the sum of the lower indices of the factors $z_j^{(i)}$ in each monomial, and we can and will always achieve this by increasing exactly one index. This amounts to replacing a $z_j^{(i)}$, which occurs in the top row of the matrix, by the element $z_{j+1}^{(i)}$ in the bottom row of the same column. This is the procedure of 'rolling factors', which can also be applied if the entries of the matrix are more general.

As example we consider the cone over $2d - b$ points in $\mathbb{P}^d$, lying on a rational normal curve of degree d, with $b < d$. Let the polynomial $P(s,t) = p_0 s^{2d-b} + p_1 s^{2d-b-1}t + \cdots + p_{2d-b}t^{2d-b}$ determine the points on the rational curve. We get the determinantal

$$\begin{vmatrix} z_0 & z_1 & \cdots & z_{d-1} \\ z_1 & z_2 & \cdots & z_d \end{vmatrix}$$

and additional equations P_m. To be specific we assume that $b = 2c$:

$$\begin{aligned} P_0 &= p_0 z_0^2 + p_1 z_0 z_1 + \cdots + p_{2d-2c-1} z_{d-c-1} z_{d-c} + p_{2d-2c} z_{d-c}^2 \\ P_1 &= p_0 z_0 z_1 + p_1 z_1^2 + \cdots + p_{2d-2c-1} z_{d-c}^2 + p_{2d-2c} z_{d-c} z_{d-c+1} \\ &\vdots \\ P_{2c} &= p_0 z_c^2 + p_1 z_c z_{c+1} + \cdots + p_{2d-2c-1} z_{d-1} z_d + p_{2d-2c} z_d^2 . \end{aligned}$$

It is not so clear what the 'generic' rolling factors singularity is. The format is not flexible: it is possible to have obstructed deformations. To describe them, we need the relations between the equations. First we note that the resolution of $\mathcal{O}_D$, where D is our divisor of type $aH - bf$, as $\mathcal{O}_S$-module is $0 \to \mathcal{O}_S(-aH+bf) \to \mathcal{O}_S \to \mathcal{O}_D \to 0$. Then we need the resolutions of these line bundles as $\mathcal{O}_\mathbb{P}$-module, described by SCHREYER [Schr]. The resolution of $\mathcal{O}_D$ is obtained by taking a mapping cone.

In general, given a map $\Phi\colon F \to G$ of locally free sheaves of rank f and g respectively, $f \geq g$, on a variety one defines Eagon-Northcott type complexes $\mathcal{C}^b$, $b \geq -1$, in the following way:

$$\mathcal{C}_j^b = \begin{cases} \bigwedge^j F \otimes S_{b-j}G , & \text{for } 0 \leq j \leq b \\ \bigwedge^{j+g-1} F \otimes D_{j-b-1}G^* \otimes \bigwedge^g G^* , & \text{for } j \geq b+1 \end{cases}$$

with differential defined by multiplication with $\Phi \in F^* \otimes G$ for $j \neq b+1$ and $\bigwedge^g \Phi \in \bigwedge^g F^* \otimes \bigwedge^g G$ for $j = b+1$ in the appropriate term of the exterior, symmetric or divided power algebra.

In our situation $F \cong \mathcal{O}_{\mathbb{P}^n}^d(-1)$ and $G \cong \mathcal{O}_{\mathbb{P}^n}^2$ with Φ given by the matrix of the scroll; it can be obtained intrinsically from the multiplication map

$$H^0 \mathcal{O}_S(R) \otimes H^0 \mathcal{O}_S(H - R) \longrightarrow H^0 \mathcal{O}_S(H) .$$

Then $\mathcal{C}^b(-a)$ is for $b \geq -1$ the minimal resolution of $\mathcal{O}_S(-aH + bR)$ as $\mathcal{O}_\mathbb{P}$-module [Schr, Cor. 1.2]. By taking the mapping cone we obtain the following first terms of the resolution:

$$0 \leftarrow \mathcal{O}_D \longleftarrow \mathcal{O}_\mathbb{P} \longleftarrow \left(\mathrm{Sym}_b\, \mathcal{O}_\mathbb{P}^2\right)(-a) \oplus \wedge^2 \mathcal{O}_\mathbb{P}^d(-1) \longleftarrow$$
$$\longleftarrow \left(\mathcal{O}_\mathbb{P}^d(-1) \otimes \mathrm{Sym}_{b-1}\, \mathcal{O}_\mathbb{P}^2\right)(-a) \oplus \wedge^3 \mathcal{O}_\mathbb{P}^d(-1) \otimes \mathcal{O}_\mathbb{P}^2 .$$

To describe the relations we introduce the following notation. A column in the matrix Φ has the form $(z_j^{(i)}, z_{j+1}^{(i)})$. We write symbolically $(z_\alpha, z_{\alpha+1})$, where the index α stands for the pair $_j^{(i)}$ and $\alpha+1$ means adding 1 to the lower index. The rolling factors assumption is that two consecutive additional equations are of the form

$$P_m = \sum_\alpha p_{\alpha,m} z_\alpha ,$$

$$P_{m+1} = \sum_\alpha p_{\alpha,m} z_{\alpha+1} .$$

where the polynomials $p_{\alpha,m}$ depend on the z-variables and the sum runs over all possible pairs $\alpha = {}_j^{(i)}$. To roll from P_{m+1} to P_{m+2} we collect the 'coefficients' in the equation P_{m+1} in a different way: we also have $P_{m+1} = \sum_\alpha p_{\alpha,m+1} z_\alpha$.

We write the scrollar equations as $f_{\alpha\beta} = z_\alpha z_{\beta+1} - z_{\alpha+1} z_\beta$. The relations between them are

$$R_{\alpha,\beta,\gamma} = f_{\alpha,\beta} z_\alpha - f_{\alpha,\gamma} z_\beta + f_{\beta,\gamma} z_\alpha ,$$
$$S_{\alpha,\beta,\gamma} = f_{\alpha,\beta} z_{\alpha+1} - f_{\alpha,\gamma} z_{\beta+1} + f_{\beta,\gamma} z_{\alpha+1} ,$$

which corresponds to the term $\wedge^3 \mathcal{O}^d(-1) \otimes \mathcal{O}^2$ in SCHREYER's resolution. The other summand yields relations involving the P_m :

$$R_{\beta,m} = P_{m+1} z_\beta - P_m z_{\beta+1} - \sum_\alpha f_{\beta,\alpha} p_{\alpha,m} ,$$

where $0 \leq m < b$. We note the following relation:

$$R_{\beta,m} z_\gamma - R_{\beta,m} z_\beta - \sum R_{\beta,\gamma,\alpha} p_{\alpha,m} = P_m f_{\beta,\gamma} - f_{\beta,\gamma} P_m .$$

The right hand side is a Koszul relation; the second factor in each product is considered as coefficient. There are similar expressions involving $z_{\gamma+1}$, $z_{\beta+1}$ and $S_{\beta,\gamma,\alpha}$.

Lemma. Let X be the cone over D. If $a > 2$, then $\dim T_X^2(-a) = b - 1$, and $\dim T_X^2(-a) \geq b - 1$ in case $a = 2$.

Proof. Let $\psi \in \mathrm{Hom}(R/R_0, \mathcal{O}_X)$ be an homogeneous element of degree $-a$. The degree of $\psi(R_{\alpha,\beta,\gamma})$ is $3 - a$, so ψ vanishes on the scrollar relations, if $a > 2$. If $a = 2$ we can assert that the functions vanishing on the scrollar relations span a subspace of $T_X^2(-2)$.

As the degree of the relation $R_{\alpha,m}$ is $a + 1$, the image $\psi(R_{\alpha,m})$ is a linear function of the coordinates. The relations

$$R_{\alpha,m}z_\beta - R_{\beta,m}z_\alpha - \sum R_{j,k,\gamma}p_{\gamma,m} = P_m f_{\alpha,\beta} - f_{\alpha,\beta}P_m.$$

imply that the $\psi(R_{\alpha,m})$ are also in rolling factors form. A basis (of the relevant subspace) of $\mathrm{Hom}(R/R_0, \mathcal{O}_X)(-a)$ consists of the $2b$ elements $\psi_{l,s}(R_{j,m})$ $= \delta_{lm}z_\alpha$, $\psi_{l,t}(R_{j,m}) = \delta_{lm}z_{\alpha+1}$, where $0 \le m < b$. The image of P_m in $\mathrm{Hom}(R/R_0, \mathcal{O}_X)(-a)$ is $\psi_{m-1,s} - \psi_{m,t}$, if $0 < m < b$, $-\psi_{0,t}$ for $m = 0$, and $\psi_{b-1,s}$ for $m = b$. The quotient has dimension $b - 1$.   □

The elements of $T^2_X(-a)$ just constructed can be considered as obstruction against rolling factors in a deformation, in which the scroll and the divisor on it are simultaneously deformed. In general a deformation of X is not the restriction of a deformation of the cone over S. On the other hand, not all deformations of the matrix extend to deformations of the divisor on the scroll. One gets linear equations on the deformation parameters, which depend on the coefficients of the polynomial P. They are described in [St9]. Given an infinitesimal lift to deformations of the divisor we can describe the deformation explicitly in case $a = 2$ [St7, Prop. 2.12]. From that one even obtains the base equations directly from the polynomial P [St9, Thm. 3.4].

Here we only continue the example of the cone over $2d - b$ points on a rational normal curve of degree d. One has $\dim T^1(\nu) = 0$ for $\nu \ne 0, -1$, $\dim T^1(0) = d(d - b - 2)$ (the number of moduli of $2d - b$ points in general position in $\mathbb{P}^d$), and $\dim T^1(-1) = 2d - b$ [St7, Prop. 2.9]. All scrollar deformations

$$\begin{vmatrix} z_0 & z_1 & \cdots & z_{d-2} & z_{d-1} \\ z_1 + \zeta_1 & z_2 + \zeta_2 & \cdots & z_{d-1} + \zeta_{d-1} & z_d \end{vmatrix}$$

lift and we have additional parameters $\rho_0, \ldots, \rho_{d-b}$. Explicit formulas for the lift are given in [St7]. As they are rather involved we write the deformation of the equations P_m only for the special case $d = 7$, $b = 5$, and $p_i = 0$ for $0 < i < 9$. Each time we go from the perturbed equation P_m to P_{m+1} by rolling factors. The condition that the expression obtained is the same as the starting point for the next step, gives the base equations.

$$p_0 z_0^2 + p_9 z_4 z_5 + \rho_0 z_0 + \rho_1 z_1 + \rho_2 z_2 - 2p_9\zeta_6 z_3 - p_9\zeta_5 z_4 ,$$

$$p_0 z_0(z_1 + \zeta_1) + p_9(z_5 + \zeta_5)(z_5 - \zeta_5) + \rho_0(z_1 + \zeta_1) + \rho_1(z_2 + \zeta_2)$$
$$+ \rho_2(z_3 + \zeta_3) - 2p_9\zeta_6(z_4 + \zeta_4)$$
$$= p_0 z_0 z_1 + p_9 z_5^2 + \rho_0 z_1 + \rho_1 z_2 + \rho_2 z_3 + p_0\zeta_1 z_0 - 2p_9\zeta_6 z_4 ,$$

$$p_0(z_1 + \zeta_1)^2 + p_9 z_5(z_6 + \zeta_6) + \rho_0(z_2 + \zeta_2) + \rho_1(z_3 + \zeta_3)$$
$$+ \rho_2(z_4 + \zeta_4) - 2p_9\zeta_6(z_5 + \zeta_5)$$
$$= p_0 z_1^2 + p_9 z_5 z_6 + \rho_0 z_2 + \rho_1 z_3 + \rho_2 z_4 + 2p_0\zeta_1 z_1 - p_9\zeta_6 z_5 ,$$

$$p_0 z_1(z_2 + \zeta_2) + p_9(z_6 + \zeta_6)(z_6 - \zeta_6) + \rho_0(z_3 + \zeta_3) + \rho_1(z_4 + \zeta_4) +$$
$$\rho_2(z_5 + \zeta_5) + 2p_0\zeta_1(z_2 + \zeta_2)$$
$$= p_0 z_1 z_2 + p_9 z_6^2 + \rho_0 z_3 + \rho_1 z_4 + \rho_2 z_5 + 2p_0\zeta_1 z_2 + p_0\zeta_2 z_1 ,$$

$$p_0(z_2 + \zeta_2)^2 + p_9 z_6 z_7 + p_0(z_4 + \zeta_4) + p_1(z_5 + \zeta_5) + p_2(z_6 + \zeta_6)$$
$$+ 2p_0\zeta_1(z_3 + \zeta_3)$$
$$= p_0 z_2^2 + p_9 z_6 z_7 + p_0 z_4 + p_1 z_5 + p_2 z_6 + 2p_0\zeta_1 z_3 + 2p_0\zeta_2 z_2 ,$$
$$p_0 z_2(z_3 + \zeta_3) + p_9 z_7^2 + p_0(z_5 + \zeta_5) + p_1(z_6 + \zeta_6) + p_2 z_6$$
$$+ 2p_0\zeta_1(z_4 + \zeta_4) + 2p_0\zeta_2(z_3 + \zeta_3) .$$

We obtain the base equations

$$\rho_0\zeta_1 + \rho_1\zeta_2 + \rho_2\zeta_3 + 2p_9\zeta_6\zeta_4 + p_9\zeta_5^2 ,$$
$$\rho_0\zeta_2 + \rho_1\zeta_3 + \rho_2\zeta_4 - p_0\zeta_1^2 + 2p_9\zeta_5\zeta_6 ,$$
$$\rho_0\zeta_3 + \rho_1\zeta_4 + \rho_2\zeta_5 - 2p_0\zeta_1\zeta_2 + p_9\zeta_6^2 ,$$
$$\rho_0\zeta_4 + \rho_1\zeta_5 + \rho_2\zeta_6 - 2p_0\zeta_1\zeta_3 - p_0\zeta_2^2 .$$

In our naive conception of format, equations play an important role; examples as the generic determinantal suggest this. In more complicated examples, as the third format for the Artin component of rational quadruple points [Rö], which we describe below, we have a rigid singularity as 'generic form', but is not quite clear what the best way is to write the equations. Therefore we split our format problem in two. We first ask:

Question. For which classes of singularities is there a preferred way to write down generators of the ideal ?

Examples are toric varieties, for which the equations are 'binomial': they express that two monomials represent the same element of the ring. Not surprisingly, toric varieties turn up in the simplest examples; maybe they suggest a too simple picture.

Not depending on equations BUCHWEITZ made the following definition [Bu]:

Definition. Let $Y \subset \mathbb{C}^N$ be a singularity. A germ $X \subset \mathbb{C}^M$ is a *singularity of type Y*, if there exists a map $\phi \colon \mathbb{C}^M \to \mathbb{C}^N$, such that $\phi^*(Y) = X$, which induces a complete intersection morphism $\phi \colon X \to Y$.

We recall that a morphism $\phi \colon X \to Y$ is a complete intersection, if it can be factored as $p \circ i$, where $i \colon X \to Y \times \mathbb{C}^k$ is a regular embedding (i.e., X is defined by a regular sequence), and $p \colon Y \times \mathbb{C}^k \to Y$ is the projection.

Deformations of type Y of X are obtained by unfolding ϕ: for every map $\Phi \colon \mathbb{C}^M \times (S,0) \to \mathbb{C}^N$, extending ϕ, the map $\pi \colon \Phi^*Y \to (S,0)$ is flat [Bu, 4.3.4]. The functor of unfoldings of ϕ is smooth, but its image in the deformation functor of X is in general not semi-homogeneous. A simple example is given by Ancus RÖHR [Rö]: consider the (non rigid) singularity $Y = A_1 \subset \mathbb{C}^4$, which we write as determinantal

$$\begin{vmatrix} x & y + s \\ y - s & z \end{vmatrix} .$$

The surface singularity $X = A_1$ is of type Y, under $\phi(x, y, z) = (x, y, z, 0)$; the given unfolding is as unfolding non-trivial in first order and even defines a non-trivial deformation of X, but it is trivial as first order deformation.

Even if the image is a smooth subspace of the base (which can be checked infinitesimally), it is not necessarily a component. Furthermore, the subspace depends on the choice of ϕ.

Example (cf. [Bu, Scha]).   The base space of the curve $X = L_4^4$ is not smooth (Chap. 3): it is the Segre cone on $\mathbb{P}^1 \times \mathbb{P}^3$, where we take $(l:m)$ and $(a:b:c:d)$ as coordinates. With coordinates (x, y, z, t) on $\mathbb{C}^4$ we can write the versal deformation of X as

$$xy - ablm$$
$$xz - clx - alz + aclm$$
$$xt - dmx - amt + adlm$$
$$yz - cmy - bmz + bclm$$
$$yt - dly - blt + bdlm$$
$$zt - dclm$$

By considering the deformation parameters as coordinates on $\mathbb{P}^1 \times \mathbb{C}^4$, we effectively make a base change to a small resolution of the base space. For fixed $(l:m)$ with $l \neq 0$, $m \neq 0$ and $l \neq m$, a maximal determinantal deformation of Y is given by

$$\begin{vmatrix} x & bl & l(z - cm) & t - dl \\ am & y & m(z - cl) & t - dm \end{vmatrix}.$$

For the exceptional values of $(l:m)$ it is impossible to write the equations in determinantal form, because for all values of the parameters a, b, c and d the deformed curve lies in $\mathbb{C}^2 \vee \mathbb{C}^2$, so has at least one singular point, whereas the general fibre of a maximal determinantal deformation of X is smooth [Scha].

We can write the equations as quasi-determinantal in a neighbourhood of the special parameter values; near $l = 0$ we get

$$\begin{vmatrix} x & t - dl & b & z - cm \\ am & t - dm & y & m(z - cl) \end{vmatrix}.$$
$$l$$

The equations of $X = L_4^4$ can also be written as 2×2 minors of a symmetric 3×3 matrix. Again, not every deformation has this format. We now take the other small resolution of the base: $\mathbb{C}^2 \times \mathbb{P}^3$. We write the equations in a kind of symmetric quasi-determinantal format; the generic singularity $\mathcal{Y}$ is given by a symmetric 3×3 matrix and a vector of multipliers for the columns or rows:

$$\begin{array}{c} \begin{array}{ccc} \alpha_1 & \alpha_2 & \alpha_3 \end{array} \\ \begin{array}{c} \alpha_1 \\ \alpha_2 \\ \alpha_3 \end{array} \begin{vmatrix} a_{11} & a_{12} & a_{13} \\ a_{12} & a_{22} & a_{23} \\ a_{13} & a_{23} & a_{33} \end{vmatrix} \end{array}.$$

We form the following generalised minors:

$$a_{ii}a_{jj} - \alpha_i\alpha_j a_{ij}^2 , \qquad i \neq j ,$$
$$a_{ii}a_{jk} - \alpha_i a_{ij}a_{ik} , \qquad i \neq j \neq k ,$$

where we use the convention $a_{ij} = a_{ji}$. One way of obtaining these expressions is the following: multiply the columns (or rows) by their multipliers except the entries on the diagonal, and divide whenever possible the minors of this matrix by the multipliers. If we assume that $d \neq 0$, we can write the versal family of L_4^4 as

$$\begin{array}{c|ccc} & a & b & c \\ \hline a & dx + ta - lad & t & t - dl \\ b & t & dy + bt - mbd & t - dm \\ c & t - dl & t - dm & dz + tc - (l+m)cd \end{array}.$$

Definition. A singularity $(X, 0)$ is a *germ without smooth factor*, if in every product decomposition of $(X, 0) \cong (Y, 0) \times (\mathbb{C}^k, 0)$ one necessarily has $k = 0$.

The terminology is admittedly arkward, but all obvious alternatives have already a different meaning. One can write any germ as product of a smooth factor and a germ without smooth factor, where the isomorphism type of the germ without smooth factor is uniquely determined, because by [Eph, Lemma 1.5] two germs are isomorphic if and only if their products with the same smooth factor are isomorphic.

Let $\pi \colon X \to S$ be the versal deformation of a singularity X_0. Suppose that $T \subset S$ is a smooth component of the base space, and let $Y \to T$ be the induced deformation over T. Decompose Y as $F \times \mathbb{C}^k$ with F a germ without smooth factor. Then X_0 is of type F. More generally, let $T \subset S$ be any component of the deformation space of X_0, and let $Y \to T$ be the corresponding deformation. The Zariski-Jacobi exact sequence for $\mathbb{C} \to \mathcal{O}_T \to \mathcal{O}_Y$ yields the exact sequence

$$T_T^0 \otimes \mathcal{O}_Y \longrightarrow T_{Y/T}^1 \xrightarrow{\ \alpha\ } T_Y^1 \longrightarrow T_T^1 \otimes \mathcal{O}_Y .$$

The proof of [Rö, 2.7.2] shows that α is the zero map, if T is a component of the base space S.

Proposition. *If T is rigid (e.g. smooth), then Y is also rigid.*

Example: the 2-star singularity. By a result of DE JONG and VAN STRATEN [JS4, Thm. 2.10] the base space of a rational quadruple point has $n+1$ irreducible components, if n is the *number of virtual quadruple points* (p. 119); more precisely, it is isomorphic to $B(n) \times S$, where $S \cong (\mathbb{C}^k, 0)$ is a smooth factor and $B(n)$ is a space of embedding dimension $5n - 1$. To define $B(n)$ one takes as coordinates the coefficients of polynomials $a_1(t)$, $a_2(t)$, $a_3(t)$ and $e(t)$ of degree $n-1$ in t, and a polynomial $b(t)$ of degree $n-2$. The conditions

$$a_i(t)e(t) \text{ is divisible by } t^n + b(t), \qquad i = 1, 2, 3,$$

determine a system of equations on the coefficients, which generate the ideal of $B(n)$.

The largest and the smallest component are smooth, but the other components are non-normal. The simplest total space for such a component is obtained for $n = 2$ from the 2-star singularity, which has the following resolution graph:

The versal deformation is described in [loc. cit.]; the equations given here were computed directly with *Macaulay*. As the modulus of the singularity gives a smooth factor, we calculated only the versal deformation in negative degree, for the value -1 of the cross-ratio. The singularity is determinantal, as is the total space over the Artin component (for the definition, see Chap. 14, Ex. 1):

$$\begin{vmatrix} x_1 & x_2 & x_3 & x_4 \\ x_2 + tb_2 + a_2 & x_3 + tb_3 + a_3 & x_4 + tb_4 + a_4 & x_1 + t^2 + e \end{vmatrix}.$$

There are two more deformation variables, which appear on the Veronese component:

$$\begin{vmatrix} x_1 & x_2 & x_3 + td + c \\ x_2 & x_3 & x_4 \\ x_3 + td + c & x_4 & x_1 + t^2 + e \end{vmatrix}.$$

The equations of the base space are

$$b_3c + a_3d, \qquad b_2c + a_2d, \qquad b_4c + a_4d,$$
$$a_3c - b_3de, \qquad a_2c - b_2de, \qquad a_4c - b_4de.$$

With a new variable u we can write the normalisation of the third component as

$$c = ud, \qquad e = -u^2,$$
$$a_2 = -ub_2, \qquad a_3 = -ub_3, \qquad a_4 = -ub_4.$$

From this description we see that the component is five dimensional, with one dimensional singular locus, and as transverse singularity the non Cohen-Macaulay singularity of two planes meeting in one point. A computation of T^1 shows that it is rigid, so the total space above it is also rigid.

The pull back of the total family to the normalisation of the component has a total space, which is again rigid, and is without smooth factor. This follows from an explicit computation of T^1. It is useful to first simplify the equations with a suitable coordinate transformation. To make the symmetry better visible, we also replace u by a new variable b_1. We get the following rigid singularity:

$$x_3^2 - x_4^2 + b_3^2 e^2 - b_4^2 e^2 - x_3 b_3 d + x_4 b_4 d \,,$$
$$x_1^2 - x_2^2 + b_1^2 e^2 - b_2^2 e^2 - x_1 b_1 d + x_2 b_2 d \,,$$
$$x_2 x_4 - x_3 b_1 e - x_1 b_3 e - b_2 b_4 e^2 + b_1 b_3 ed \,,$$
$$x_2 x_3 - x_4 b_1 e - x_1 b_4 e - b_2 b_3 e^2 + b_1 b_4 ed \,,$$
$$x_1 x_4 - x_3 b_2 e - x_2 b_3 e - b_1 b_4 e^2 + b_2 b_3 ed \,,$$
$$x_1 x_3 - x_4 b_2 e - x_2 b_4 e - b_1 b_3 e^2 + b_2 b_4 ed \,.$$

Example: a non-rigid component. The previous example shows that in many cases, even if a component is not smooth, it is still rigid. I computed the base space for a special rational singularity of multiplicity 6, which occurs in [Kol, 6.3.6]. This singularity has a component, over which there is a one-dimensional family of P-modifications (see Chap. 14). The resolution graph is

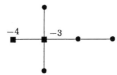

The computations are rather long. I give here only the result for the deformation in negative degree for the special value -1 of the cross-ratio. The variables are $a, \ldots, g$, while the deformation parameters come directly from my computation, and have the names x_2, x_3, x_6, x_{10} and x_{11}. After some simplifications I found the following equations for the total space of the component in question:

$$cb - da$$
$$(d - ax_3 - (b - 2x_{10})x_2)(b - x_{10}) - ea$$
$$c^2 + f(a + x_{11}) - (b - 2x_{10} + x_2^2)a^2$$
$$ec - ga + (b - x_{10})a(a + x_6 + x_{11}) + (b - x_{10})^2 x_{10}$$
$$-dc - fb + (b - x_{10})^2 a + (b - 2x_{10})ax_2^2 - a^2 x_{11} + cx_3 x_{11}$$
$$e(d + x_2 x_{10}) - (g - b(a + x_6 + x_{11}))(b - x_{10})$$
$$fb - a^2(a + x_6) + (c + ax_2)ax_3$$
$$-b(b - x_{10})^2 + (g + ex_2)a$$
$$-(d + x_2 x_{10})(d - ax_3 - (b - 2x_{10})x_2) + ga - ba(a + x_6 + x_{11})$$
$$fd - ca(a + x_6) + (c + ax_2)cx_3$$
$$(g + ex_2)c - d(b - x_{10})^2$$

$$-fe + (b - x_{10})((c - ax_2)(a + x_6) - (b - 2x_{10})ax_3)$$
$$(g + ex_2)(d - ax_3 - (b - 2x_{10})x_2) - eb(b - x_{10})$$
$$(g + ex_2)f - (b - x_{10})^2(a(a + x_6) - (c + ax_2)x_3)$$
$$-g^2 + e^2(b + x_2^2) + (g + ex_2)b(a + x_6 + x_{11}) + eb(b - x_{10})x_3 \ ,$$

while the base space is given by the determinantal

$$\begin{vmatrix} 2x_2 & -x_{11} & x_{10} \\ x_3 & x_{10} & x_6 \end{vmatrix}.$$

This is not the full component, as we have fixed the modulus. However this will not change the structure of the base space. It is not rigid, but has a one-dimensional T^1, which corresponds to making the matrix generic. This base space has a small resolution with exceptional fibre $\mathbb{P}^1$, and the small resolution is the base space for the deformations of the P-modifications. The total space is also not rigid — this is a *Macaulay* computation of 333 minutes on an IBM RS 6000 workstation.

In his Hamburg thesis [Rö] Ancus RÖHR turns the problem of formats around, and defines a *format* for the Artin component of a rational surface singularity as the result of cancelling smooth factors of the total space: write the total space $(Y, 0)$ over the Artin component as $(F \times \mathbb{C}^k, 0)$ with k maximal, and define the space F to be the format. He proves that for rational quadruple points three formats exist, the determinantal (the total space over the Artin component of the cone over the rational normal curve), the quasi-determinantal (the total space for the cyclic quotient $X_{8,5}$), and a third one, of which I write out the six equations; they have a lot of structure, but it seems that there is no compact way of giving them:

$$x_1 x_4 - x_2 x_3 + y_1 y_4 - y_2 y_3$$
$$x_1 y_2 - x_2 y_1 - z_2(x_3 y_2 - x_4 y_1) + z_3(x_3 y_4 - x_4 y_3)$$
$$x_1 y_4 - x_2 y_3 - z_1(x_3 y_2 - x_4 y_1) + z_2(x_3 y_4 - x_4 y_3)$$
$$x_1^2 + (z_1 z_3 - z_2^2)x_3^2 - z_1 y_1^2 + 2z_2 y_1 y_3 - z_3 y_3^2$$
$$x_1 x_2 + (z_1 z_3 - z_2^2)x_3 x_4 - z_1 y_1 y_2 + z_2(y_1 y_4 + y_2 y_3) - z_3 y_3 y_4$$
$$x_2^2 + (z_1 z_3 - z_2^2)x_4^2 - z_1 y_2^2 + 2z_2 y_2 y_4 - z_3 y_4^2 \ .$$

Note that the last three equations describe quadratic forms with discriminant $(z_1 z_3 - z_2)^2$. Blowing up a canonical ideal gives a small modification with exceptional locus in $F \times \mathbb{P}^2$ over $x_i = y_i = 0$, given by $t_2^2 = z_1 t_1^2 + 2z_2 t_1 t_3 + z_3 t_3^2$.

13 Smoothing components of curves

If the fibre X_s over some point s in the base space of a versal deformation $X \to S$ is smooth, then by openness of versality all nearby fibres are also smooth and s is a regular point of S. In the terminology of WAHL [Wa2] the irreducible component of S, containing s, is a *smoothing component*. The first question arising is of course the existence of smoothing components. The second asks for their dimension. Here the results depend on the dimension of the space X_0. In high dimension many interesting singularities are rigid. The problem, whether rigid reduced curve singularities, or rigid normal surface singularities exist, is still open.

For a reduced curve singularity the dimension e of smoothing components is an invariant of the singularity, which does not depend on the component; it can be computed with a formula of DELIGNE, which simplifies in the quasi-homogeneous case to $e = \mu + t - 1$ [Greu], with μ the Milnor number and t the Cohen-Macaulay type. For Gorenstein singularities $t = 1$, and in general t is the number of generators of the dualising module. For curves with embedded components the Milnor number does depend on the component [BG], and the dimension of the smoothing component also. In [St6, Sect. 2] the versal deformation of the general hyperplane section of two planes meeting in a point is computed. This curve, consisting of two crossing lines with an embedded component sticking out in a new direction, can also be smoothed by smoothing the normal crossing and moving the point away; such deformations make up a component of dimension 4. This example suggests that in general the difference in dimension of smoothing components is three times the difference in Milnor number. In the surface case the dimension is already variable for normal singularities (the standard example is the cone over the rational normal curve of degree four), with difference in dimension twice the difference in Milnor number: there is a formula of the form $e = d(X_0) + 2\alpha(\text{cpt})$ [Wa2], where $d(X_0)$ is an invariant of the singularity only, and α is related to the Milnor number over the component. If the surface singularity is Gorenstein ($t = 1$), then the Milnor number, and with it e, is an invariant of the singularity. One dimension higher even Gorenstein singularities have components of different dimension; e.g., the cone over a Del Pezzo surface of degree 6 has as base space a plane and a line meeting transversally. One may speculate about additional properties, which force threefold singularities to have components of the same dimension, and then about the four dimensional case. Often minimally elliptic surface singularities are regarded as the Gorenstein counterpart of rational surface singularities (the calculus of cycles on the exceptional divisor is similar in both cases). The remarkably resembling features of the three examples of increasing dimension above suggest that three dimensional (rational) Gorenstein singularities are the more natural generalisation.

We now concentrate on smoothings of reduced curve singularities. We consider a very simple type of singularity, a curve L^n_r, consisting of r general lines through the origin in $\mathbb{C}^n$, or equivalently, the affine cone over r general points in $\mathbb{P}^{n-1}$. Under certain conditions every deformation X_t of $X_0 = L^n_r$ comes from a projective deformation $\overline{X}_t$ of $\overline{X}_0$, the projective cone over the r points. If $\overline{X}_t$ is smooth, then $\overline{X}_{t,\infty}$ is a hyperplane section of a smooth curve and $\overline{X}_{0,\infty}$, our set of r points, is at least a limit of hyperplane sections. By results of PINKHAM and GREUEL [Greu, Pin1] the conditions are satisfied for the curve L^n_r, which is the cone over a set of r points in generic position in $\mathbb{P}^{n-1}$, with $n < r \le \binom{n+1}{2}$. Therefore this curve has smooth deformations if and only if the set of r points lies in the closure of the locus of hyperplane sections; the genus of the smooth curve is $g = r - n$. In [St1] I proved:

Proposition. Let $n < r \le \binom{n+1}{2}$. The general L^n_r has smooth deformations if and only if $(r - n - 2)(n - 5) \le 6\varepsilon$ for $r - n \ne 8$, where $\varepsilon = 0$ or 1, according to the parity of $r - n$.

The proof is based on the relation between hyperplane sections of non-special curves and point sets on canonical curves. The following concept seems to occur in this generality for the first time with COBLE [Cob]. It is also known as Gale transform, see [EP]. We keep COBLE's terminology.

Definition. Let $\Pi = (p_1, \ldots, p_r)$ be a set of r ordered points in $\mathbb{P}^{g-1}$, whose coordinates are given by a $g \times r$-matrix P. A set of r points $\{q_1, \ldots, q_r\}$ in $\mathbb{P}^{r-g-1}$, given by a $(r - g) \times r$-matrix Q, is called *associated* to Π if $PQ^t = 0$.

Lemma. If no subset consisting of $r - 1$ points of $\Pi = (p_1, \ldots, p_r) \in (\mathbb{P}^{g-1})^r$ is contained in a hyperplane, then the associated set in $(\mathbb{P}^{r-g-1})^r$ is uniquely determined up to projective transformations.

Proof. Let the matrix P represent Π. We may suppose that P is of the form (I_g, A). Write $Q = (Q_1, Q_2)$ with Q_1 a $(r - g) \times g$-matrix. Then $PQ^t = Q_1^t + AQ_2^t = 0$ implies that $Q = (-A^t, I)Q_2$. Because every $r - 1$ points span $\mathbb{P}^{g-1}$, no row of A is zero, so $(-A^t, I)$ determines r points in $\mathbb{P}^{r-g-1}$. This association is independent of the choice of P, because the expression $PQ^t = 0$ is invariant under the action $(A, B, C; P, Q) \mapsto (APB^{-1}, CQB^t)$ of $Gl(g) \times Gl(r) \times Gl(r - g)$ on $M(g, r) \times M(r - g, r)$. $\square$

Proposition. Let $\{p_1, \ldots, p_r\}$ be a hyperplane section of a non-special curve C of genus g in $\mathbb{P}^{r-g}$. Let $\phi_K : C \to \mathbb{P}^{g-1}$ be the canonical embedding. Then $\{p_1, \ldots, p_r\}$ and $\{\phi(p_1), \ldots, \phi(p_r)\}$ are associated point sets.

Proof. Let D be the hyperplane section. We have the exact sequence

$$0 \to H^0(\mathcal{O}) \longrightarrow H^0(\mathcal{O}(D)) \xrightarrow{A^t} H^0(\mathcal{O}_D(D)) \longrightarrow H^1(\mathcal{O}) \to 0 \,,$$

where A^t is the matrix of the map $f \mapsto \{f(p_i)\}$. We may take a generator of the image of $H^0(\mathcal{O})$ as first basis vector of $H^0(\mathcal{O}_D(D))$, so A^t has the form $(0, A_1^t)$ for a $r \times (r-g)$-matrix A_1^t, whose transpose A_1 describes the points p_i in $\mathbb{P}^{r-g-1}$. We compare the Serre dual of this sequence with the exact sequence in which the map $B^t \colon \phi \mapsto \{\phi(p_i)\}$ occurs:

$$
\begin{array}{ccccccccc}
0 & \to & H^0(\omega) & \xrightarrow{B^t} & H^0(\omega \otimes \mathcal{O}_D) & \longrightarrow & H^1(\omega(-D)) & \longrightarrow & H^1(\omega) & \to & 0 \\
 & & \downarrow{\scriptstyle\cong} & & \downarrow{\scriptstyle\cong} & & \downarrow{\scriptstyle\cong} & & \downarrow{\scriptstyle\cong} & & \\
0 & \to & H^1(\mathcal{O})^* & \longrightarrow & H^0(\mathcal{O}_D(D))^* & \xrightarrow{A} & H^0(\mathcal{O}(D))^* & \longrightarrow & H^0(\mathcal{O})^* & \to & 0
\end{array}
$$

This shows that $AB^t = 0$ and therefore $A_1 B^t = 0$. $\square$

This brings us to the question through how many general points in $\mathbb{P}^{g-1}$ passes a canonical curve of genus g. More generally we define

$$N(n, d, g) = \sup \{\, r \mid \text{ through } r \text{ general points in } \mathbb{P}^n \text{ passes a } {}^g C_1^d \,\}; ,$$

where ${}^g C_1^d$ stands for a smooth curve of genus g and degree d. The case $n = 3$ was studied by PERRIN [Per]. Such problems can be studied by infinitesimal methods.

Let $H(p)$ be the Hilbert scheme of closed subschemes of $\mathbb{P}^n$ with Hilbert-Samuel polynomial $p(z) = r$, and $H(q)$ the Hilbert scheme of the polynomial $q(z) = dz + 1 - g$. Let H_r be the open subscheme of $H(p)$, which contains the subschemes D consisting of r distinct points of $\mathbb{P}^n$, and let $H_{d,g}$ be the open subscheme of $H(q)$ whose rational points are the smooth connected curves C of degree d and genus g in $\mathbb{P}^n$. Consider also the flag scheme $F(p, q)$: an S-valued point is a couple (X_S, Y_S) of closed subschemes $X_S \subset Y_S \subset \mathbb{P}^n_S$, flat over S, where for $s \in S$ the fibres X_s and Y_s have Hilbert-Samuel polynomials p and q. The scheme $F(p, q)$ has two natural projections: $f \colon F(p, q) \to H(p)$ and $g \colon F(p, q) \to H(q)$. In $F(p, q)$ we consider $F_{r;d,g} = f^{-1}(H_r) \cap g^{-1}(H_{d,g})$; we denote the restrictions of the projections f and g by f_r and g_r. Then

$$N(n, d, g) = \sup \{\, r \mid \text{im } f_r \text{ contains a non-empty open subset of } H_r \,\} \,.$$

We look at these maps at the level of tangent spaces. The Zariski tangent space to $H(p)$ in a point X is $H^0(X, N_X)$. We have the following general result of KLEPPE [Per]:

Theorem. *Let (X, Y) be a rational point of $F(p, q)$ and let T be the Zariski tangent space to $F(p, q)$ in (X, Y). Then the diagram*

$$
\begin{array}{ccc}
T & \xrightarrow{dg} & H^0(Y, N_Y) \\
\downarrow{\scriptstyle df} & & \downarrow{\scriptstyle r} \\
H^0(X, N_X) & \xrightarrow{s} & H^0(X, N_Y|_X)
\end{array}
$$

is Cartesian. Suppose furthermore that $H(p)$ is smooth in Y. If the map $r : H^0(Y, N_Y) \to H^0(X, N_Y|_X)$ is surjective, then f is smooth in the point (X, Y). In particular, the image of f contains an open neighbourhood of X.

In the special case that X is an effective divisor D on a smooth curve C the restriction map r sits in the cohomology sequence of the exact sequence $0 \to N_C(-D) \to N_C \to N_C \otimes \mathcal{O}_D \to 0$. So to prove that $N(n, d, g) \geq r_0$ it suffices to find a curve C and a divisor D, consisting of r_0 distinct points, with $H^1(N_C(-D)) = 0$. In fact one only needs such a D on a singular curve C in the closure of $H_{d,g}$: then $H^1(N_C) = 0$, so $H(q)$ is smooth in C, df is surjective, and f is smooth in (C, D).

In the case of canonical curves we have:

Lemma. *The number $r(g) = N(g - 1, 2g - 2, g)$ satisfies $r(g) \leq g + 5 + [6/(g - 2)]$.*

Proof. The scheme H_r is smooth of dimension $r(g - 1)$, while $F_{r;2g-2,g}$ is smooth of dimension $3g - 3 + (g^2 - 1) + r$, so when f_r is surjective, then $r + g^2 + 3g - 4 \geq r(g - 2)$.   □

In [St1] I constructed a curve C with g cusps, and a Weil divisor D, consisting of the g cusps and 5 other points, such that $H^1(N_C(-D)) = 0$. This shows that $r(g) \geq g + 5$. It remains to look at low values of g:

$g = 3$: plane quartics are determined by 14 points, $r(3) = 14$.

$g = 4$: a canonical curve is the complete intersection of a cubic and a unique quadric. The quadric is determined by 9 points, while 6 additional points on it specify the curve: $r(4) = 9$.

$g = 5$: the general curve is a complete intersection of three quadrics, so $r(5) = 12$.

$g = 6$: the general curve of genus 6 has a plane representation as a curve of degree 6 with 4 double points, so the canonical curve lies on a Del Pezzo surface S of degree 5. In the exact sequence $0 \to N_{C/S} \to N_C \to N_S|_C \to 0$ the degree of N_C is 70, and $\deg N_{C/S} = 20$. For a general divisor D of degree 12 $H^1(N_{C/S}) = 0$; by Riemann-Roch $h^1(N_C(-D)) = h^1(N_S|_C(-D)) \geq 1$, so the map f_{12} cannot be surjective.

$g = 7$: an explicit example of a reducible curve with 13 points on it, given in [St1], shows $r(7) = 13$.

In these examples $r(g) = g + 5$ for g even and $r(g) = g + 5 + [6/(g - 2)]$ for odd g. This suggests for the remaining case $r(8) = 13$. Despite great efforts I did not succeed in proving this. In fact, it turns out that $r(8) = 14$. The problem in computations is that one needs a general curve. The original problem turned out to be easier.

Proposition. *A general curve singularity L_{14}^6 is smoothable, and has 16 smoothing components of dimension 36, while $\dim T^1 = 43$.*

Proof. The statement is true for the ideal of 14 lines in $(\mathbb{Z}_{31991})^6$: I computed it with *Macaulay* [BS]. A discussion of the computation can be found in [St6].   □

The curve L_{14}^6 gives to my knowledge the first example of a curve singularity with several smoothing components. As the moduli of the curve enter in the equations of the base space, it is hopeless to find irreducible components and to actually write down a smoothing. But it is possible to compute with a special L_{14}^6, which is still 'general enough'. The first idea, to take the 6 coordinate axes and the cone over 8 points on a rational normal curve, does not work, because the ideal is then generated by quadrics and cubics. Instead, consider 6 coordinate points, 6 points $s^6 - t^6$ on the curve $(s^6, s^5t, \ldots, t^6)$, and two other points; after some experimentation I took $(1,1,0,1,0,1)$ and $(1,0,1,0,1,1)$. The first 12 points give a L_{12}^6 with particular nice equations:

$$x_i x_{i+3} - x_{i+1} x_{i+2}, \quad x_{i-1} x_{i+1} - x_{i-2} x_{i+2}, \quad i \in \mathbb{Z}/6\mathbb{Z}.$$

This curve has only equisingular deformations: T^1 is concentrated in degree 0. This shows again that $r(6) = 11$.

The points $(1,1,0,1,0,1)$ and $(1,0,1,0,1,1)$ impose two conditions on the nine given quadrics. The dimension of T_{-1}^1 of the resulting singularity is 8, as in the generic case. I computed the versal deformation in negative degree. After some coordinate transformations I found the following equations for the base space S:

$$s_2 s_1, \quad s_3 s_4, \quad s_2 s_3, \quad s_2 s_8, \quad s_3 s_5, \quad s_1 s_5, \quad s_4 s_8,$$

$$(3s_2 + s_5)s_7, \quad (3s_3 + s_8)s_6, \quad s_2(s_5 + 3s_6), \quad s_3(s_8 + 3s_7),$$

$$s_2(6s_4 - 2s_6 + 3s_7), \quad s_3(6s_1 - 2s_7 + 3s_6),$$

$$s_4(6s_4 + 3s_7 - 2s_6 - 4s_1), \quad s_1(6s_1 + 3s_6 - 2s_7 - 4s_4),$$

$$(2s_1 - 2s_3 + s_6)s_8, \quad (2s_4 - 2s_2 + s_7)s_5,$$

$$(2s_2 - 2s_4 - s_7 + s_8)s_6, \quad (2s_3 - 2s_1 - s_6 + s_5)s_7,$$

$$s_1 s_4 - s_6 s_7, \quad s_5 s_7 + s_6 s_8 + 3s_6 s_7.$$

The equations for the total space are still rather complicated. The $\mathbb{Z}/2\mathbb{Z}$-symmetry on the singularity induces the involution

$$(s_1, \ldots, s_8) \mapsto (s_4, s_3, s_2, s_1, s_8, s_7, s_6, s_5)$$

on the base space S. There are nine components. Three of them have dimension two, C_7: the (s_7, s_8)-plane, C_8: the (s_5, s_8)-plane, and C_9: the (s_5, s_6)-plane; C_8 intersects C_7 and C_9 in a line, while $C_7 \cap C_9 = \emptyset$. The other components have dimension one; four of them are reduced, while C_5 and C_6 have a multiplicity two structure. The general fibre over each of these components is reducible.

Each line L in S defines a one parameter deformation $\mathcal{X} \to L$ of X; all fibres $\mathcal{X}_l$ with $l \neq 0$ are isomorphic, and the projective closure of $\mathcal{X}_l$ is isomorphic to the projective curve, defined in $\mathbb{P}^7 = \mathbb{P}(\mathbb{C}^6 \times L)$ by the equations of $\mathcal{X}$.

The total space over the component C_8 is given by the equations:

$$x_3 x_4 - x_2 x_5$$
$$x_2 x_3 - x_4 x_5 + x_3 x_6 - x_5 x_6$$
$$-x_1 x_2 + x_1 x_4 + x_3 x_6 - x_5 x_6$$
$$x_1 x_2 - x_2 x_3 + x_2 x_5 - x_4 x_5 - x_1 x_6 + x_5 x_6$$
$$x_1 x_2 + x_1 x_3 - x_3 x_6 - x_4 x_6 - x_3 s_8 + x_4 s_5$$
$$x_1 x_2 + x_3 x_5 - x_2 x_6 - x_3 x_6 - x_1 s_8 + x_2 s_5$$
$$-x_1 x_4 + x_2 x_4 - x_1 x_5 + x_5 x_6 + x_5 s_8 - x_6 s_5 \ .$$

It has seven components of degree two, given by:

$$(x_4 - x_6, x_3 - x_5, x_2 - x_6, x_5 + s_8, (x_1 + x_5)x_5 + (x_1 - x_5 - x_6 + s_5)x_6) \ ,$$
$$(x_1 - x_3, x_2 - x_4, x_1 - x_5, x_2 + s_5, x_2(x_2 + x_6) + x_1(x_6 - x_1 - x_2 + s_8)) \ ,$$
$$(x_1, x_3, x_5, x_6 - s_5, x_2 x_4 - x_6^2) \ ,$$
$$(x_2, x_4, x_6, x_1 - s_8, x_3 x_5 - x_1^2) \ ,$$
$$(x_4 - \omega x_6, x_2 - \omega^2 x_6, x_3 - \omega^2 x_1, x_5 - \omega x_1, x_1(x_1 - s_8) - \omega^2 x_6(x_6 - s_5)) \ ,$$

with $\omega^3 = 1$ (giving three quadrics). Over the intersection with the component C_7 one has $s_5 = 0$, so the second and third quadric are reducible. The projective curve now consists of four lines:

$$i_1 = (x_1, x_2, x_3, x_4, x_5) \ ,$$
$$i_2 = (x_1 - x_3, x_2, x_4, x_1 - x_5, x_6 - x_1 + s_8) \ ,$$
$$i_3 = (x_1, x_2, x_3, x_5, x_6) \ ,$$
$$i_4 = (x_1, x_3, x_4, x_5, x_6) \ ,$$

and five conics:

$$i_5 = (x_4 - x_6, x_3 - x_5, x_2 - x_6, x_5 + s_8, (x_1 - x_6)(x_5 + x_6) + x_5^2) \ ,$$
$$i_6 = (x_1 - s_8, x_2, x_4, x_6, x_1^2 - x_3 x_5) \ ,$$
$$i_7, i_8, i_9 = (x_4 - \omega x_6, x_2 - \omega^2 x_6, x_3 - \omega^2 x_1, x_5 - \omega x_1, x_1(x_1 - s_8) - \omega^2 x_6^2) \ .$$

Through the point $P = (0 : 0 : 0 : 0 : 0 : 0 : 1)$ the lines i_1, i_2, i_3, and all conics pass. In affine coordinates the lines are the coordinate axes in the three dimensional (x_2, x_4, x_6)-space, while the tangent to a conic is given by $x_2 = \omega^2 x_6$, $x_4 = \omega x_6$; this are six lines, not on a quadric cone, so P is a singular point of type L_6^3. The intersection $(0 : -1 : 0 : -1 : 1 : 0 : 1)$ of i_4, i_6 and i_7 is of type L_3^3. Finally there are 6 ordinary double points: i_3 intersects i_4, i_6 intersects i_5, i_8 and i_9, while i_5 intersects i_7 in two points. The curve has indeed arithmetic genus 8.

The embedding dimension of the singular points is at most 3. This implies that the base space of the versal deformation (in all degrees) is smooth in a general point of $C_7 \cap C_8$. The same is true over $C_8 \cap C_9$. Therefore these three components of S lie in the same smoothing component. This shows that one has to be careful, when drawing conclusion from the negative degree part about the versal base. What happens here can be understood in the following way: the three components together are the cone over a curve of degree 3 of arithmetic genus 0; the ideal of such a curve is generated by 3 quadrics, yoked together by 2 linear relations. A general (non flat) perturbation of the quadrics gives a complete intersection, consisting of 8 points, and if the number of deformation parameters is large enough, the total space will be smooth. The base space is not normally flat along the equisingular locus.

The double component C_5 has ideal $(s_8, s_5 + 3s_6, 2s_4 - s_1 + s_7, s_3, s_1 + 2s_6, s_2 - s_6, s_6^2)$, so its reduction is the line $2s_4 + s_7 = 0$ in the (s_4, s_7)-plane. The total space over this line has 7 components; it is the cone over a curve with one singular point of type L_3^3, and 12 ordinary double points. Therefore this curve represents also a smooth point of the base of the versal deformation. A model for the smoothing component in question is a pinch point times a smooth factor. The same holds for the other double component of S. In this way we have accounted for the 16 smoothing components of the general L_{14}^6.

The equations of the special L_{14}^6 contain relatively few monomials. In practice this is a necessary condition to obtain useful explicit results. Only special, very symmetric equations are suitable for computation — however they can be representative for the general case.

14 Kollár's conjectures

Rational surface singularities are smoothable and their base space has a distinguished smoothing component, called the Artin component, which is obtained by deforming the minimal resolution [Art2]. For the special case of quotient singularities KOLLÁR and SHEPHERD-BARRON explained the reduced components of the versal deformation with so-called P-resolutions [KSh]. KOLLÁR conjectured a generalisation to all rational singularities, and introduced P-modifications [Kol]. Examples show that the same ideas might work also for non-rational singularities.

Let X_0 be a normal surface singularity, and X the total space of a one parameter smoothing of X_0. The canonical sheaf ω_X is a reflexive sheaf, but its tensor powers are in general not reflexive. Therefore one defines: $\omega_X^{[n]} = (\omega_X^{\otimes n})^{**}$, for $n > 1$. For negative n we set $\omega_X^{[n]} = \operatorname{Hom}_X(\omega_X^{[-n]}, \mathcal{O}_X)$. These sheaves are divisorial sheaves, and we can also write $\omega_X^{[n]} = \mathcal{O}_X(nK_X)$, see [Re1].

If the *canonical algebra*

$$R(X, K_X) = \sum_{n=0}^{\infty} \mathcal{O}_X(nK_X)$$

is a finitely generated $\mathcal{O}_X$-algebra, then $Y = \operatorname{Proj} R(X, K_X)$ has several interesting properties: $\pi \colon Y \to X$ is a small modification (i.e., the exceptional locus has codimension at least two), some multiple of K_Y is Cartier and K_Y is π-ample. The definition of P-modification is designed to make $Y_0 = \pi^{-1}(X_0)$ a P-modification of X_0; it should be characteristic for the component. KOLLÁR conjectures that $R(X, K_X)$ is finitely generated, whenever X_0 is a rational singularity [Kol, 6.2.1].

Definition [Kol, 6.2.3, 6.2.6]. Let $(S_0, 0)$ be a reduced (not necessarily isolated) Cohen- Macaulay surface singularity, such that $S_0 \backslash 0$ is Gorenstein. Let S be the total space of a one parameter smoothing S_t of S_0. The smoothing is $\mathbb{Q}$-*Gorenstein*, or qG, if some multiple of the canonical class of S is Cartier. If S_0 has an isolated singularity (so $\dim T^1 < \infty$), a smoothing component of the versal deformation of S_0 is called a qG-component if one (or every) smoothing on it is qG.

As KOLLÁR remarks, it would be nicer to have a functorial definition of a qG-family. It is however not clear how the embedded components of the base enter the picture. Therefore the definition involves components of the reduced base. For rational singularities every component is a smoothing component; a general surface singularity may not be smoothable at all.

If $X \to D$ is a qG-smoothing of X_0, then the index r of X is the same as the index of X_0, by [Kol, 6.2.4]. This does not imply that X is a quotient of a deformation of the canonical cover $(X_0)^c$ of X_0: the canonical cover X^c is a normal 3-dimensional singularity, and the fibre $(X^c)_0$ is isomorphic to $(X_0)^c$, if and only if X^c is Cohen-Macaulay. Because $\mathcal{O}_{X^c} = \oplus_{i=0}^{r-1} \omega_X^{[i]}$, the condition is that the natural map

$$\omega_X^{[k]} \otimes \mathcal{O}_{X_0} \longrightarrow \omega_{X_0}^{[k]}$$

is surjective for all k (up to $r-1$). For $k = 0, 1$ this map is surjective, so qG deformations of index two are quotients of deformations of the canonical cover. I do not know an example where X^c is not Cohen-Macaulay. In [JS5] it is proved that a one parameter deformation of an index r singularity is qG if and only if $\omega_X^{[1-r]} \otimes \mathcal{O}_{X_0} \longrightarrow \omega_{X_0}^{[1-r]}$ is surjective.

Maybe one should add the condition that X^c is Cohen-Macaulay to the definition of a qG-smoothing.

Definition [Kol, 6.2.10'.1]. Let X be a normal surface singularity, and let $\pi \colon Y \to X$ be a proper modification. Then Y is called a *P-modification* if

(i) $R^1 \pi_* \mathcal{O}_Y = 0$,
(ii) K_Y is π-ample,
(iii) Y has a smoothing which induces a qG smoothing of each singularity of Y.

Remark. The condition (i) guarantees that deformations of Y blow down to deformations of X [Rie2].

We now formulate a conjecture, which is more or less equivalent to Kollár's conjecture [Kol, 6.2.18] for rational singularities. We do not really expect it to be true for all normal surface singularities, but we do not know the correct class of singularities, for which one should make the conjecture.

Conjecture. Let $\pi \colon \mathcal{X} \to S$ be the total family over a smoothing component of the versal deformation of the normal surface singularity X. Then there exists a proper modification $\mathcal{Y}$ of $\mathcal{X}$ with Stein factorisation $\mathcal{Y} \xrightarrow{\pi'} T \xrightarrow{\sigma} S$, such that some multiple of $K_{\mathcal{Y}}$ is Cartier, and $K_{\mathcal{Y}}$ is π'-ample; π' is flat, and every fibre $\mathcal{Y}_t$ over $t \in \sigma^{-1}(0)$ is a P-modification of X, and the germ (T, t) is a component of the versal deformation of $\mathcal{Y}_t$.

The conjecture is true in several cases:

Example 0: Gorenstein surface singularities. In this case the conjecture is trivially true. All components have the same dimension, but they can be of different type. For example, a simple elliptic singularity X of multiplicity 8 has 5 components, four of which are smooth, and permuted by automorphisms of X, while the fifth component is singular: it is isomorphic to the singularity itself [St8, (4.4)]. The rational singularity with a (-6)-curve as central curve and four (-2)-curves is the $\mathbb{Z}/2$-quotient of X; it has three qG-components of dimension 1, which come from three of the four smooth components of the base of X.

Example 1: The Artin component.

Definition. Let $f: X \to S$ be a deformation of a singularity X_0. A *simultaneous resolution* is a proper map $\pi: \widetilde{X} \to X$, such that $f \circ \pi$ is flat, and for every $s \in S$ the fibre $\widetilde{X}_s$ is a resolution of X_s. Likewise a map $\pi: \widetilde{X} \to X$ is a *simultaneous canonical model*, if every $\widetilde{X}_s$ is the canonical model (also called RDP-resolution) of X_s.

M. ARTIN [Art2] proved that the base space of a versal deformation of a rational surface singularity has a unique irreducible component A (the *Artin component*), such that every deformation with simultaneous resolution is a pull-back from A. Conversely, every deformation, which is a pull-back from A, has after finite base change a simultaneous resolution. In particular, the total space over the Artin component has simultaneous resolution after base change; the resolved family is the versal deformation of the minimal resolution. The base change is entirely due to rational double point configurations on the minimal resolution: the Artin component is versal for deformations with simultaneous canonical model, and its canonical model is the versal deformation of the canonical model $\widehat{X}_0$ of X_0 [Wa1, Li]. Because $\widehat{X}_0$ has only isolated singularities and $H^2(\widehat{X}_0, \mathcal{F})$ vanishes for every coherent sheaf $\mathcal{F}$, the deformations of $\widehat{X}_0$ are unobstructed and the Artin component A is smooth. One constructs the simultaneous canonical model by blowing up a canonical ideal of the total space.

The dimension of the Artin component is maximal among all components; by [Wa1, 3.18] the dimension of a component is $h^1(\Theta_{\widetilde{X}_0}) - 2(q(X_0) - \alpha)$, where $\widetilde{X}_0$ is the minimal resolution and $q(X_0) = h^1(\widetilde{X}_0, \Omega^*)$. Here only α depends on the component, and $0 \leq \alpha \leq q(X_0)$. The dimension of the Artin component is $h^1(\Theta_{\widetilde{X}_0})$ (as this is the dimension of the deformation space of the minimal resolution). In all examples I know the Artin component is the only one of dimension $h^1(\Theta_{\widetilde{X}_0})$, and all other components have smaller dimension.

Example 2: Quotient singularities. By [KSh] the *canonical model* Y of a 1-parameter smoothing X of a quotient X_0 gives a small modification. The fibre Y_0 is a P-modification of X_0, which is normal, and has as singularities only rational double points and singularities of type T, which are special cyclic quotient singularities. This construction gives a $(1-1)$-correspondence between P-modifications and components for quotient singularities [KSh].

A cyclic quotient singularity X is of type T, if its canonical cover is a A_{m-1}-singularity: this happens if and only $i_\varepsilon = j_\varepsilon$ (notation as on p. 73) for some index ε, so $z_\varepsilon = (xy)^{i_\varepsilon}$. Then $n = (a_\varepsilon - 1)i_\varepsilon^2$ and $m = (a_\varepsilon - 1)i_\varepsilon$, and X has a smoothing with Milnor number $a_\varepsilon - 2$, which is the quotient of a smoothing of A_{m-1}. In terms of the resolution graph we have the following inductive characterisation:

Lemma.

(i) The singularities $\overset{-4}{\blacksquare}$ and $\blacksquare\!\!-\!\!\overset{2}{\bullet}\!\!-\cdots-\!\!\overset{2}{\bullet}\!\!-\!\!\blacksquare$ are of type type T,

(ii) If $\overset{-b_1}{\bullet}\!\!-\cdots-\!\!\overset{-b_r}{\bullet}$ is of type T, then also $\bullet\!\!-\!\!\overset{-b_1}{\bullet}\!\!-\cdots-\!\!\overset{-(b_r+1)}{\bullet}$.

In [St2] I determined all P-modifications for cyclic quotient singularities. It was observed by Jan CHRISTOPHERSEN that the components of the base space can be described by continued fractions, which represent zero (cf. [Chr, St2]). Let K_{e-2} be the set of all $[\boldsymbol{k}] = [k_2, \ldots, k_{e-1}]$ with $k_i \in \mathbb{N}$, such that $[\boldsymbol{k}] = 0$. The number of elements of K_{e-2} is the Catalan number $\frac{1}{e-2}\binom{2(e-3)}{e-3}$ [St2, 1.3]. The main result of [St2] is:

Theorem. Let X be a cyclic quotient singularity with $[\boldsymbol{a}] = [a_2, \ldots, a_{e-1}]$, and let $K_{e-2}(X) = \{[\boldsymbol{k}] \in K_{e-2} \mid k_i \leq a_i$ for all $i\}$. The reduced base S_{red} of the versal deformation of X has exactly $\#K_{e-2}(X)$ irreducible components. Each component is smooth.

The deformation over each component can be written down explicitly. Its total space is (after a coordinate transformation) a toric variety. In fact, the P-modifications can be obtained directly by toric methods [Alt2]. To describe a toric modification we have to give a fan in the first quadrant in $\mathbb{R}^2 \cong N \otimes \mathbb{R}$, where N is the lattice $\mathbb{Z}^2 + \mathbb{Z} \cdot \frac{1}{n}(1, q)$. It is well known that the minimal resolution is obtained by drawing a ray through each lattice point on the boundary of $\sigma \setminus 0$ with σ the intersection of N with the first quadrant. The canonical model is obtained by taking only those rays for which the corresponding b_i in the continued fraction satisfies $b_i > 2$. This means that the Newton polygon has a corner there. Alternatively we can look at the dual cone $\sigma^\vee$ in the dual lattice M of the exponents of invariant Laurent monomials. The rays with $b_i > 2$ are perpendicular to the sides of the Newton polygon with as vertices the interior generators $(i_\varepsilon, j_\varepsilon)$, $\varepsilon = 2, \ldots, e - 1$.

For a continued fraction $[k_2, \ldots, k_{e-1}] = 0$ the solutions to the equations $q_{i+1} + q_{i-1} = k_i q_i$ with initial conditions $q_0 = 0$, $q_1 = 1$ and with $q_e = 0$, $q_{e-1} = 1$ coincide. The set of numbers $(q_2, \ldots, q_{e-2})$ determines $[k_2, \ldots, k_{e-2}] = 0$. We now form a new Newton diagram in $M \otimes \mathbb{R}$ with vertices $(i_\varepsilon/q_\varepsilon, j_\varepsilon/q_\varepsilon)$, $\varepsilon = 2, \ldots, e - 1$. It defines a fan in the first quadrant $N \otimes \mathbb{R}$, whose rays are perpendicular to the sides of this Newton polygon. In particular, for a singularity of type T we obtain a polygon consisting of two lines parallel to the coordinate axes, so the fan has no interior rays and indeed there should not be a blow-up.

The P-modifications for the other quotient singularities were determined in [St4].

In most cases it is easier to find the P-modifications first. Contrary to the case of quotient singularities it is in general unknown what singularities can occur on P-modifications. KOLLÁR's conjectures give a nice explanation for components, but the problem of determining the number of components from the P-modifications is very difficult. If the number of components is known for other reasons, one can check the conjectures by finding P-modifications. Rather wild phenomena occur, as shown by the examples in [Kol]: non-normal P-modifications and one parameter families.

Let $Y_0 \to X_0$ be a P-modification of a rational singularity, and let $Y \to T$ be a component of the versal deformation of Y_0. There is a blowing down map $T \to S$ to the base space of the versal deformation of X_0. This map is not proper, if there exists a positive dimensional family of P-modifications.

Definition [Kol, 6.2.11]. A P-modification $Y_0 \to X_0$ is *weakly rigid*, if $Y_t \cong Y_0$ for every 1-parameter deformation $Y \to D$, such that Y_t is a P-modification of X_0 for all $t \in D$.

The qG-components of the deformation space of a weakly rigid P-modification are expected to map properly and surjectively on components of the deformation space of X_0; the functor $\mathbf{P}\text{-}mod(X/S)$ of [Kol, 6.2.13], which generalises the functor Res of [Art2], should play an important role in the proof.

In [St3] I proved a weaker result, that the blowing down map for normal, equivariant weakly rigid P-modifications maps components to smoothing components of the same dimension. Recall that a modification $\pi : Z \to X$ is called equivariant, if $\pi_* \Theta_Z = \Theta_X$. The minimal resolution of a surface singularity is equivariant [BW, Prop. (1.2)]. The proof given there, establishes that the normalised blow-up $Bl_p Y$ is equivariant, if p is a point on an equivariant normal partial resolution $Y \to X$, at which every global section of Θ_Y vanishes; such points include the singular points of Y, but also the singular points of the exceptional divisor. Together with a result of KOLLÁR, that the blowing down map is one-to-one over smoothings, this yields that it is the normalisation, if the qG-components of the singularities of Y_0 are smooth. No example of non smooth qG-components is known.

Example 3: Rational quadruple points. We first describe the rational quadruple points with qG-components. In [St3] I proved that these are exactly the ones with $2K$ Cartier. The proof of both directions used the more or less explicit classification of rational quadruple points in that paper, although for the necessary condition $2K$ integral this could be avoided. Later on DE JONG and VAN STRATEN used their projection method to prove the existence of qG smoothings in a simpler way [JS5]. Here we give a direct argument avoiding the classification.

If $2K$ is integral, then the canonical cover X^c of X is $\mathrm{Specan}(\mathcal{O}_X \oplus \omega_X)$. Over $\mathcal{O}_X$ the algebra $\mathcal{O}_X \oplus \omega_X$ has four generators, say 1, z_1, z_2 and z_3. The covering involution acts by $z_i \mapsto -z_i$. The products $z_i z_j$ lie in the maximal ideal of $\mathcal{O}_X$, so are functions in the generators x_i of $\mathfrak{m}_X$. In order to say more about these products we first prove:

Lemma. *Let X be a rational singularity. The number $q(X)$ is equal to the δ-invariant of a general canonical divisor.*

Proof. By duality on the minimal resolution $\widetilde{X}$ we have

$$q(X) = \dim H^1(\widetilde{X}, \Omega^*) = \dim H^1_E(\widetilde{X}, \Omega^{\otimes 2}) \ .$$

Let $\widetilde{D}$ be a general canonical divisor on $\widetilde{X}$, which intersects the exceptional set E transversally; its image D is a general canonical divisor of X. Consider the commutative diagram

$$
\begin{array}{ccccccccc}
 & & 0 & & 0 & & & & \\
 & & \downarrow & & \downarrow & & & & \\
0 & \longrightarrow & H^0(\widetilde{X}, \Omega_{\widetilde{X}}) & \longrightarrow & H^0(X, \omega_X) & \longrightarrow & 0 & & \\
 & & \downarrow & & \downarrow & & \downarrow & & \\
0 & \longrightarrow & H^0(\widetilde{X}, \Omega_{\widetilde{X}}^{\otimes 2}) & \longrightarrow & H^0(X, \omega_X^{[2]}) & \longrightarrow & H^1_E(\widetilde{X}, \Omega_{\widetilde{X}}^{\otimes 2}) & \longrightarrow & 0 \\
 & & \downarrow & & \downarrow & & \downarrow & & \\
0 & \longrightarrow & H^0(\widetilde{D}, \Omega_{\widetilde{D}}) & \longrightarrow & H^0(D, \omega_D) & \longrightarrow & \omega_D/\Omega_{\widetilde{D}} & \longrightarrow & 0 \\
 & & \downarrow & & \downarrow & & \downarrow & & \\
 & & 0 & & 0 & & 0 & &
\end{array}
$$

The dimension of $\omega_D/\Omega_{\widetilde{D}}$ equals $\delta(D)$. $\qquad\square$

In the case of rational singularities the map $H^0(\widetilde{X}, \Omega_{\widetilde{X}}) \otimes H^0(\widetilde{X}, \Omega_{\widetilde{X}}) \to H^0(\widetilde{X}, \Omega_{\widetilde{X}}^{\otimes 2})$ is surjective. One way to see this is with the Base Point Free Pencil Trick (see [ACGH, p. 126], the proof works also in our case). Take a two-dimensional subspace $V \subset H^0(\widetilde{X}, \Omega_{\widetilde{X}})$, spanned by sections without zeroes in common, and consider the exact sequence

$$0 \longrightarrow \mathcal{O}_{\widetilde{X}} \longrightarrow V \otimes_{\mathbb{C}} \Omega_{\widetilde{X}} \longrightarrow \Omega_{\widetilde{X}}^{\otimes 2} \longrightarrow 0 \ .$$

By rationality $V \otimes H^0(\widetilde{X}, \Omega_{\widetilde{X}}) \to H^0(\widetilde{X}, \Omega_{\widetilde{X}}^{\otimes 2})$ is surjective.

So, if $2K$ is integral, the ring $\mathcal{O}_X/(z_i z_j)$ is as $\mathcal{O}_D$-module isomorphic to $\omega_D/\Omega_{\widetilde{D}}$. For a rational quadruple point a general canonical divisor is a curve of type A_{2n-1} or A_{2n}; here n is the *number of virtual quadruple points*, which is defined by the condition that the embedding dimension drops only after n blow-ups, or alternatively it is the maximal number of quadruple points, into which the singularity deforms. It follows that $\mathcal{O}_X/(z_i z_j)$ has embedding dimension at most one; if $n > 1$, the embedding dimension is exactly one. Therefore four (or five) linear independent linear forms in the coordinates x_i of X enter in the expressions $z_i z_j$, and can be eliminated. So the embedding dimension of X^c is 3 if $n = 1$, and 4 otherwise. As there are 6 products $z_i z_j$, we can conclude in the case $n > 1$ that the two equations f_1, f_2 of X^c have a quadratic part, which involves some of these products; therefore they can be chosen to be invariant under the covering transformation. By taking $f_1 = at$, $f_2 = bt$ for generic a, b, we obtain an involution invariant smoothing. If $n = 1$, the canonical cover is a double point.

Corollary. *A rational quadruple point with $2K$ integral is a symmetric determinantal singularity.*

Proof. Consider the symmetric matrix $(z_i z_j)$, with the entries written as functions of the coordinates x_i. The minors of this matrix vanish on X. Using the matrix we can construct a double cover of its zero set, which coincides with the canonical cover X^c of X. Therefore the minors define X. □

A rational quadruple point with n as number of virtual quadruple points has a base space with $n + 1$ irreducible components [JS4]. The resolution graph of X contains a maximal subgraph, on which $2K$ is integral [St3]: it is the support of $\lceil \lceil 2K \rceil \rceil$, where $\lceil \lceil 2K \rceil \rceil$ is the smallest integral cycle Y on the exceptional set E satisfying $E_i \cdot Y \leq E_i \cdot 2K$ for all irreducible components E_i. A P-modification is obtained from the minimal resolution by blowing down the support of $\lceil \lceil 2K \rceil \rceil$ and all maximal rational double point configurations, which do not intersect $\lceil \lceil 2K \rceil \rceil$. There are $n - 1$ subcycles of $\lceil \lceil 2K \rceil \rceil$, on which $2K$ is also integral: they are the singularities, obtained by blowing up the singularity with $2K$ integral. From each subgraph with $2K$ integral we obtain a P-modification; the canonical model gives the $(n+1)$-st component, the Artin component.

Example 4: Determinantal quadruple points. Our first set of examples for non rational singularities comes from determinantal singularities with equations

$$X: \quad \begin{vmatrix} z_1 & z_2 & z_3 & z_4 \\ z_2 & z_3 & z_4 & f(z_1, s) \end{vmatrix},$$

with $f(0,0) = 0$ and $f(0,s) \not\equiv 0$. The base space of these singularities is the same as that of rational quadruple points by [JS4], because the multiplicity of the double curve Σ of a generic projection to $\mathbb{C}^3$ is three; one way to compute this is to look at a general hyperplane section. It suffices to take $s - cz_1$ for some constant c; the resulting curve is isomorphic to the determinantal curve

$$Y: \quad \begin{vmatrix} z_1 & z_2 & z_3 & z_4 \\ z_2 & z_3 & z_4 & z_1^m \end{vmatrix},$$

where m is the multiplicity of f. The projection on the (z_1, z_2)-plane is $z_2^4 = z_1^{3+m}$. The multiplicity of Σ is equal to the difference in δ-invariant between the curve section and its plane projection (see p. 87), which is easily computed.

Every singularity X of this type has a qG-component. We can write the equations also as the 2×2 minors of a symmetric 3×3 matrix, so X is the $\mathbb{Z}/2$-quotient of the complete intersection in $\mathbb{C}^4$

$$y^2 = xz , \qquad z^2 = f(x^2, s) .$$

The case of rational quadruple points suggest that to find other P-modifications one has to blow up. If $m = 1$, then either X is the cone over the rational normal curve of degree 4, or we can bring f to the form $z_1 + s^n$, so X is the n-star of [JS4]. If the multiplicity of f is at least 2, then the reduced tangent cone of X is given by $z_2 = z_3 = z_4 = 0$. Therefore two coordinate charts suffice to cover the exceptional curve E of the blow-up $\pi\colon X_1 \to X$. In $(z_1, \zeta_2, \zeta_3, \zeta_4, \sigma)$-coordinates, where $z_i = z_1 \zeta_i$ and $s = z_1 \sigma$, we can eliminate ζ_3 and ζ_4, and the strict transform X_1 of X is given by $\zeta_2^4 = z_1^{-1} f(z_1, z_1 \sigma)$. In the (ζ_i, s)-chart, where $z_i = s\zeta_i$, the space X_1 has a singularity of multiplicity 4:

$$\begin{vmatrix} \zeta_1 & \zeta_2 & \zeta_3 & \zeta_4 \\ \zeta_2 & \zeta_3 & \zeta_4 & s^{-1} f(s\zeta_1, s) \end{vmatrix} .$$

This is a singularity of the same type we are considering, so it has a qG-component. A direct computation gives that $R^1 \pi_* \mathcal{O}_{X_1} = 0$. To check the third condition for a P-modification, we write down a global section of $2K$. On the double cover of the quadruple point on X_1 the 2-form $\omega = dx \wedge ds/yz$ generates K; we express its square in our two coordinate patches:

$$\omega^2 = \frac{(d\zeta_1 \wedge ds)^2}{s^{-1} f(\zeta_1 s, s) \zeta_2^2} = \sigma \frac{(dz_1 \wedge d\sigma)^2}{\zeta_2^6} = \sigma(\omega')^2 ,$$

where ω' generates the canonical module at the origin in the first coordinate patch. This shows that $K \cdot E > 0$.

To obtain the P-modification, belonging to the next bigger component, we blow up once again. The strict transform E_1 of E is covered by two coordinate charts, (z_1, ζ_2, σ) and $(z_1', \zeta_2', \sigma')$ with $(z_1, z_2, s) = (z_1, z_1\zeta_2, z_1\sigma) = ((z_1')^2\sigma', (z_1')^2\sigma'\zeta_2', z_1'\sigma')$, and the equations of the surface are

$$\zeta_2^4 = z_1^{-1} f(z_1, z_1\sigma) , \qquad (\zeta_2')^4 = (z_1'\sigma')^{-1} f((z_1')^2\sigma', z_1'\sigma') .$$

This shows that E_1 can be blown down, and the strict transform X_2 of X is then locally a fourfold branched cover of an A_1-singularity. In general, the $(k + 1)$-th P-modification is obtained by blowing up k times, has one exceptional curve and one singularity which is the fourfold branched cover of an A_{k-1}. This process continues until the embedding dimension goes down.

There is a second method to obtain the 'Artin component', i.e. the biggest one, namely by blowing up a canonical ideal. In this case this is the same as a so called Tyurina modification [Str]: the ratio between the rows of the defining matrix gives a rational map to $\mathbb{P}^1$, and the closure of its graph gives the wanted modification. We get from homogeneous coordinates (u, v) on $\mathbb{P}^1$ two coordinate charts (z_1, s, v) and (z_4, s, u), and the strict transform $\widehat{X}$ is given by $f(z_1, s) = v^4 z_1$, and $z_4 = f(u^3 z_4, s)$; in the second chart the origin is always a smooth point. This modification has at most hypersurface singularities, giving a quick way to find the minimal resolution of a P-modification.

We illustrate the above by some more specific examples.

(a) Let $f = z_1^2 + s^8$. Then X is the quasicone $X(C, L)$ with C a hyperelliptic curve of genus 3 (the curve $w^2 = x^8 + 1$) and $L = \mathcal{O}_C(p)$, where p is not a Weierstraß point. On the Tyurina modification we have the singularity $z_1^2 + s^8 = v^4 z_1$, the quasicone with $L = g_2^1$: the modification can be described on the minimal resolution as blowing up the image of the point p under the hyperelliptic involution on the exceptional divisor. By blowing up once we obtain $\zeta_2^4 = z_1 + \sigma^8 z_1^7$; after 8 blow ups we can blow down to $\zeta_2^4 = z_1 + t$, $s^8 = z_1 t$; elimination of t gives the same equation as before.

(b) Let $f = z_1^2 s + s^3$. By blowing up once we obtain a non-normal P-modification X_1, which in the s-chart has the determinantal description

$$\begin{vmatrix} z_1 & z_2 & z_3 & z_4 \\ z_2 & z_3 & z_4 & (z_1^2 + 1)s^2 \end{vmatrix}.$$

To find the correct number of P-modifications it is essential that we only blow up, and do not normalise: normalising X_1 gives a surface with an elliptic curve as exceptional set and four A_1-points. The singularity at the origin in the s-chart is isomorphic to a determinantal with $f = s^2$; blowing up this point gives as new quadruple point the cone over the rational normal curve.

Example 5. Let X be the cone over a hyperelliptic curve C of genus 3, embedded with a linear system L of degree 8; the curve C is the intersection of a quadric and a scroll $S(a, b)$ with $a + b = 4$. If $L = 2K$, then the scroll is the projective cone over a rational normal curve of degree 4, and the quadric can be given as $t^2 = Q(z_0, \ldots, z_4)$. Explicit smoothings are described in [Te3]. One can show that for $L \neq 2K$ there are two smoothing components, but three in case $L = 2K$. Then the singularity has a qG-component: the canonical cover is the quasicone $X(C, K)$, the complete intersection $x_0 x_2 - x_1^2$, $y^2 = q_4(x)$ (where q_4 is a polynomial of degree 4 in the x_i, obtained from the polynomial $q_8(s, t)$ defining the Weierstraß points). The $\mathbb{Z}/2$-action is multiplication of the x_i by -1. For the cone over the bicanonical embedding of a non hyperelliptic curve with $g = 3$ there are two components, the smallest obtained from even degree deformations of a homogeneous degree 4 hypersurface singularity under the Veronese embedding of $\mathbb{C}^3$; in particular the dimension of the component is 14, in agreement with the results of TENDIAN.

The P-modification of the largest component of the base of X is obtained by blowing up a canonical ideal; this amounts to making the Tyurina modification for the affine cone over the projective cone over the rational normal curve of degree 4 (i.e. $X_{4,1} \times \mathbb{C}^1$), on which X lies. The resulting surface has an exceptional rational curve with transverse A_1-singularities and 8 pinch points.

The middle component has the most interesting P-modifications. We first note that a general degree zero deformation leads to the cone over an non hyperelliptic curve C, embedded with a general linear L system of degree 8. A smoothing of such a curve is easily obtained: take 8 points $P_1, \ldots, P_8$ such that $L + P_1 + \ldots + P_8 = 4K$ and embed $\mathbb{P}^2$, blown up in the 8 points P_i on the canonical image of C, with the linear system of quartics with the P_i as base points; sweeping out the cone over this surface defines the required smoothing.

Let more generally X be the cone over $S = \varphi_L(\mathbb{P}^2)$, where L is the linear system of quartics with k base points in general position. We want generators for the canonical ring $R(X, K_X)$. One has

$$\mathcal{O}_X(nK_X) = \bigoplus_{m \in \mathbb{Z}} H^0(\mathcal{O}_S(nK_S + mL)) \, ,$$

and $K_S = -3H + \sum_i E_i$, $L = 4H - \sum_i E_i$ with the E_i the exceptional curves over the points P_i. In particular, $H^0(\mathcal{O}_S(4K_S+3L)) = \mathbb{C}$, and $H^0(\mathcal{O}_S(4K_S+4L)) = \mathbb{C}[x,y,z]_4$ (the polynomials of degree 4), and these spaces generate $\sum \mathcal{O}_X(4nK)$.

The Proj of the canonical ring has the fourfold Veronese embedding of $\mathbb{C}^3$ as singular point. We give two descriptions of this space Y. The surface S is obtained by projecting the fourfold Veronese embedding of $\mathbb{P}^2$ from the span of k points; likewise X is obtained by a linear projection $\mathbb{C}^k \times \mathbb{C}^{15-k} \to \mathbb{C}^{15-k}$, and Y is the closure of X in $\mathbb{P}^k \times \mathbb{C}^{15-k} \to \mathbb{C}^{15-k}$. The second description is in terms of flops. Blowing up the cone X gives an exceptional surface, isomorphic to S, containing k rational curves E_i with normal bundle $\mathcal{O}(-1) \oplus \mathcal{O}(-1)$; flopping these curves gives as exceptional set $\mathbb{P}^2$ and 8 transversal lines. Blowing down $\mathbb{P}^2$ gives Y.

Now let the curve C of genus 3 be embedded with $|L| = |4K - \sum P_i|$. The cone $X(C, L)$ over C has a P-modification Y with one singular point of type $X(C, 4K)$ and k exceptional curves, passing through the singular point; their tangent lines are linearly independent. The points P_i are not uniquely determined by L; in fact, every point set of the linear system $|4K - L|$ will do. We obtain in this way a $\mathbb{P}^{k-3}$ of P-modifications; if $\sum n_i P_i \in |4K - L|$, then the construction is as follows: blow up n_i times in P_i on the exceptional divisor of the minimal resolution of $X(C, L)$, and blow down C (which has now normal bundle $-4K$) and the A_{n_i-1} configurations. This construction works also if C is hyperelliptic and if $L = 2K$.

We remark that for the cone over a general curve of degree 8 and genus 3 the dimension of T^1 is 15, with $\dim T^1_0 = 9$ and $\dim T^1_{-1} = 6$. Apparently the map $T \to S$ from the conjecture is blowing up the equisingular locus (i.e. the degree zero subspace). Furthermore, already the non-uniqueness of the P-modifications on this component shows that one does not obtain a P-modification by blowing up a canonical ideal; this can be interpreted as the non existence of an 'Artin component'.

Example 6. Finally we describe an interesting example of a rational singularity, due to WAHL [Letter to Kollár, 1988], where a qG-component exists only for a special value of the modulus of a singularity. The resolution graph is

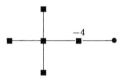

Here a square ■ without self-intersection number stands for a (-3)-curve. The canonical covering has degree 6, and is an elliptic singularity of multiplicity 18, with $p_g = 2$. The graph is

For special values of the moduli this singularity is itself a $\mathbb{Z}/21\mathbb{Z}$ quotient of a hypersurface singularity: consider the cone over the sextic curve $f_6 = x^5 y + y^5 z + z^5 x$. On it acts a non abelian group of order 126, with generators

$$\alpha\colon (x,y,z) \mapsto (\zeta x, \zeta^2 y, \zeta^4 z) \,,$$
$$\beta\colon (x,y,z) \mapsto (-\nu y, -\nu^4 z, -\nu^7 x) \,,$$

where ζ is a primitive seventh root of unity, and ν a primitive ninth root of unity. The quotient is the rational singularity.

We form the quotient in several stages. The group of order 3, generated by β^6, acts diagonally by multiplying with third roots of unity, so the quotient of $\mathbb{C}^3$ under this action is the cone over the 3-fold Veronese embedding of $\mathbb{P}^2$, which induces the canonical embedding of the sextic curve. The dimension of a smoothing component of the cone over a canonical curve of genus g is equal to $2g+21$ (by the formulas of [Wa2]), so 41 for $g = 10$; 41 is also the dimension of the invariant subspace of the T^1 of the cone over the curve f_6. Therefore the invariant deformations of f_6 are a component of the deformation space of the corresponding canonical cone. The dimension of T^1 is 49, and the parts of different degree have the following dimension:

$k =$	-2	-1	0	1	2
$\dim T^1(k) =$	1	10	27	10	1

As the general canonical cone of genus 10 is smoothable, there are at least two components for curves with a g_6^2.

The next step is to divide out by the $\mathbb{Z}/7\mathbb{Z}$-action, generated by α. The deformations of f_6 invariant under β^6 and α are

$$x^5y + y^5z + z^5x + \lambda_0 + \lambda_1 xyz + \lambda_2 (xyz)^2 + \lambda_3 (xyz)^3 + \lambda_4 (xyz)^4 \ .$$

We get a 5-dimensional smoothing component. The modulus of the elliptic curve varies on the component. The normal bundle of the elliptic curve in the minimal resolution has as divisor the intersection points with the three rational curves. The positions of these points give at least two extra moduli, which are fixed on the component. To study these points, it suffices to consider the $\mathbb{Z}/7\mathbb{Z}$-action on the sextic curve. Invariants are $(a : b : c) = (xy^3 : yz^3 : zx^3)$, which gives the elliptic curve $E : a^2b + b^2c + c^2a$. The coordinate triangle is at the same time an inscribed and a circumscribed triangle; let its vertices be P, Q and R. If we take P as base point of the group law on E, then $Q + R = 0$, $2Q = R$, so Q and R are points of order three.

The canonical cover of a rational singularity with resolution graph as above, is always an elliptic singularity as considered, with the special property that $j = 0$ for the central elliptic curve. The intersection points with the rational curves are three points, which are a divisor in the g_3^1. Only if the cross-ratio on the central rational curve is $1/2 \pm 1/2\sqrt{3}i$ (giving $j = 0$), we get collinear flexes as above. It gives a one dimensional smoothing component with Milnor number zero.

It seems that there are six of these smoothing components, which correspond to taking a different primitive ninth root of unity ν in the definition of β. If we multiply ν with a third root of unity, the $\mathbb{Z}/21\mathbb{Z}$-quotient of f_6 is unchanged. We can lift the isomorphism between the quotients for different ν to an automorphism of the elliptic singularity. On the elliptic curve $E : a^2b + b^2c + c^2a$ the Cremona transformation $(a : b : c) \mapsto (ab : ac : cb)$ acts as involution, which transforms the action of β, given by $(a : b : c) \mapsto (\nu^3b : \nu^6c : a)$ into $(a : b : c) \mapsto (c : \nu^3a : \nu^6b)$, which is the inverse of $(a : b : c) \mapsto (\nu^6b : \nu^3c : a)$. It is not difficult to compute the induced involution on the elliptic singularity — it acts non-trivially on the smoothing. By symmetry we have two more Cremona transformations. Together they generate a group of order six. A general principle may be involved: different but 'similar' smoothing components of a Gorenstein surface singularity are related by automorphisms of the singularity.

15 Cones over curves

Cones over smooth projective varieties provide a simple class of isolated singularities. Their deformation theory was studied at an early stage. The basic reference is a paper by SCHLESSINGER [Schl2] and MUMFORD's 'footnote' to it [Mu3]. We start with the description of Chapter 10 for T_X^1 of a normal singularity $(X, 0) \subset (\mathbb{C}^n, 0)$. With $U = X \setminus \operatorname{Sing} X$ we have

$$T_X^1 = \operatorname{coker}\{H^0(U, \Theta_n|_X) \longrightarrow H^0(U, N_U)\} ,$$

or

$$T_X^1 = \ker\{H^1(U, \Theta_U) \longrightarrow H^1(U, \Theta_n|_X)\} .$$

In the special case that X is the cone over a projective variety Y the space U is a $\mathbb{C}^*$-bundle over Y and we want to interpret the occurring cohomology groups on Y. From now on we work in the algebraic category.

We start from a smooth projective variety Y with a specific embedding: let L be a very ample line bundle on Y and let $\phi_L : Y \to \mathbb{P}(V^*)$ be the corresponding embedding, where $V = H^0(Y, L)$. We will identify Y and $\phi_L(Y)$ and also write $L = \mathcal{O}_Y(1)$. Our singularity $X \subset V^*$ is the the affine cone over Y. We suppose that X is normal, or in other words that $\phi_L(Y)$ is projectively normal. The smooth space $U := X \setminus 0$ is a $\mathbb{C}^*$-bundle over Y. We denote the projections $\pi : U \to Y$ and $\pi : V^* \setminus 0 \to \mathbb{P}(V^*)$ by the same symbol π. If $\mathcal{F}$ is a sheaf on X with a natural $\mathbb{C}^*$-action, then $\pi_* \mathcal{F}$ decomposes into direct sums of the eigenspaces for the various characters of $\mathbb{C}^*$. Let $\mathcal{F}_Y$ be the sheaf of $\mathbb{C}^*$ invariants. Then

$$H^0(X, \mathcal{F}) = H^0(U, \mathcal{F}) = \bigoplus_{\nu=-\infty}^{\infty} H^0(Y, \mathcal{F}_Y(\nu)) .$$

The last equality does not hold in the analytic context; in that case one has a filtration on $H^0(X, \mathcal{F})$, whose associated graded space is the direct sum above.

When explicitly writing down sections of sheaves, it is more convenient to use homogeneous coordinates: this really means to compute on X instead of Y. We illustrate this with the tangent bundle of $\mathbb{P}^n$. One has the Euler sequence [Ha, II.8.20.1]

$$0 \longrightarrow \mathcal{O}_{\mathbb{P}^n} \longrightarrow \mathcal{O}_{\mathbb{P}^n}(1)^{n+1} \longrightarrow \Theta_{\mathbb{P}^n} \longrightarrow 0 .$$

In homogeneous coordinates $(z_0, \ldots, z_n)$ elements of a basis of $H^0(\mathcal{O}_{\mathbb{P}^n}(1)^{n+1})$ can be written as $z_i \frac{\partial}{\partial z_j}$, and the map from $\mathcal{O}_{\mathbb{P}^n}$ is given by $1 \mapsto \sum_i z_i \frac{\partial}{\partial z_i}$.

Next we recall the definition of the *sheaf of principal parts* [Kl, IV.A], [EGA, IV.16.7]. Consider a scheme Y; let $\mathcal{J}$ be the ideal sheaf of the diagonal Δ in $Y \times Y$, and let $Y_\Delta^{(n)}$ be the n-th infinitesimal neighbourhood of Δ. The canonical projections p_1 and p_2 of the product induce maps $p_1^{(n)}: Y_\Delta^{(n)} \to Y$ and $p_2^{(n)}: Y_\Delta^{(n)} \to Y$. We define for a sheaf $\mathcal{F}$ [EGA, IV.16.7.2]

$$\mathcal{P}_Y^n(\mathcal{F}) = (p_1^{(n)})_* \left((p_2^{(n)})^*(\mathcal{F}) \right) .$$

We write $\mathcal{P}_Y^n$ for $\mathcal{P}_Y^n(\mathcal{O}_Y)$. One has $\mathcal{P}_Y^n(\mathcal{F}) = \mathcal{P}_Y^n \otimes_{\mathcal{O}_Y} \mathcal{F}$, where the tensor product is taken with the $\mathcal{O}_Y$-module structure, defined by p_2. Because the diagonal is a section of $Y \times Y$ for both p_1 and p_2, both morphisms define a homomorphism $\mathcal{O}_Y \to \mathcal{P}_Y^n$, and therefore an $\mathcal{O}_Y$-module structure. Except when explicitly stated, we always consider the $\mathcal{O}_Y$-module structure on $\mathcal{P}_Y^n$, induced by p_1, and write it as left multiplication. One denotes by d^n the morphism $\mathcal{O}_Y \to \mathcal{P}_Y^n$, induced by p_2 [EGA, IV.16.3.6]. For every $t \in \Gamma(U, \mathcal{O}_Y)$, $U \subset Y$ open, $d^n t$ is the *principal part of order n*. In particular, $dt = d^1 t - t \in \Gamma(U, \Omega_Y^1)$ is the *differential* of t. We have the exact sequence

$$0 \longrightarrow \Omega_Y^1 \longrightarrow \mathcal{P}_Y^1 \longrightarrow \mathcal{O}_Y \longrightarrow 0 .$$

On $\mathbb{P}^n$ this sequence is the dual of the Euler sequence.

The sheaf $\mathcal{P}_Y^n(\mathcal{F})$ also has two $\mathcal{O}_Y$-module structures; it is convenient to write them on the left and the right. For $a \in \Gamma(U, \mathcal{O}_Y)$, $b \in \Gamma(U, \mathcal{P}_Y^n)$ and $t \in \Gamma(U, \mathcal{F})$ one has [EGA, iV.16.7.4]

$$a(b \otimes t) = (ab) \otimes t ,$$
$$(b \otimes t)a = (b \cdot d^n a) \otimes t = b \otimes (at) = (d^n a) \cdot (b \otimes t) .$$

There is a map $d_\mathcal{F}^n: \mathcal{F} \to \mathcal{P}_Y^n(\mathcal{F})$ with $d_\mathcal{F}^n(t) = 1 \otimes t$.

Definition [EGA, IV.16.8.1]. Let $\mathcal{F}$ and $\mathcal{G}$ be two $\mathcal{O}_Y$-modules. A homomorphism $D: \mathcal{F} \to \mathcal{G}$ is a *differential operator of order* $\leq n$ if there exists a homomorphism $u: \mathcal{P}_Y^n(\mathcal{F}) \to \mathcal{G}$ such that $D = u \circ d_\mathcal{F}^n$.

The differential operators form a group; by applying the construction on open sets we obtain a sheaf $\mathcal{D}\mathit{iff}_Y^n(\mathcal{F}, \mathcal{G})$, which is isomorphic to the sheaf $\mathcal{H}om_{\mathcal{O}_Y}(\mathcal{P}_Y^n(\mathcal{F}), \mathcal{G})$ [EGA, IV.16.8.4]. We write $\mathcal{D}\mathit{iff}_Y^n$ for $\mathcal{D}\mathit{iff}_Y^n(\mathcal{O}_Y, \mathcal{O}_Y)$, and $\mathcal{D}\mathit{iff}_Y^n(\mathcal{F})$ for $\mathcal{D}\mathit{iff}_Y^n(\mathcal{F}, \mathcal{O}_Y)$.

We return to our map $\phi_L: Y \to \mathbb{P}(V^*)$. Then $\phi_L^*(\mathcal{P}_{\mathbb{P}(V^*)}^1) = V \otimes_\mathbb{C} L^{-1}$. Here V is considered to be the vector space with dz_i as basis. Let N_Y^* be the conormal bundle of Y in $\mathbb{P}(V^*)$. Because Y is smooth, one has the familiar exact sequence $0 \longrightarrow N_Y^* \longrightarrow \phi_L^*(\Omega_{\mathbb{P}(V^*)}^1) \longrightarrow \Omega_Y^1 \longrightarrow 0$, with which we obtain the following result [Kl, (IV.19)]:

Proposition. *In the situation as above the following sequence is exact:*

$$0 \longrightarrow N_Y^* \otimes L \longrightarrow V \otimes_{\mathbb{C}} \mathcal{O}_Y \longrightarrow \mathcal{P}_Y^1(L) \longrightarrow 0 \,,$$

or dually:

$$0 \longrightarrow \mathcal{D}iff_Y^1(L) \longrightarrow V^* \otimes_{\mathbb{C}} \mathcal{O}_Y \longrightarrow N_Y \otimes L^{-1} \longrightarrow 0 \,.$$

In particular we can view the $\mathbb{C}^*$-invariants of the exact sequence, defining T_X^1, as obtained by taking global sections of the sequence $0 \longrightarrow \mathcal{D}iff_Y^1 \longrightarrow V^* \otimes_{\mathbb{C}} L \longrightarrow N_Y \longrightarrow 0$ on Y. We get the following formulation of a result of [Schl2]:

Theorem. *Let L be a very ample line bundle on a smooth projective variety Y, which embeds Y as projectively normal subvariety of $\mathbb{P}(V^*)$, where $V = H^0(Y, L)$. Let $X \subset V^*$ be the affine cone over Y. Then the graded parts $T_X^1(\nu)$ of T_X^1 are given by*

$$T_X^1(\nu) = \operatorname{coker}\left\{ V^* \otimes H^0(Y, L^{\nu+1}) \longrightarrow H^0(Y, N_Y \otimes L^\nu) \right\},$$

or alternatively,

$$T_X^1(\nu) = \ker\left\{ H^1(Y, \mathcal{D}iff_Y^1 \otimes L^\nu) \longrightarrow V^* \otimes H^1(Y, L^{\nu+1}) \right\}.$$

From now one we concentrate on the case that Y is a curve, and write C for Y; Serre duality then transforms the second formula into one involving H^0.

Corollary. *Let X be the cone over a curve C, embedded by $L = \mathcal{O}_C(1)$. Then*

$$(T_X^1)^*(\nu) = \operatorname{coker}\left\{ V \otimes H^0(C, \Omega_C^1(-\nu-1)) \longrightarrow H^0(C, \mathcal{P}_C^1 \otimes \Omega_C^1(-\nu)) \right\}.$$

This Corollary makes it possible to determine the graded parts of T^1 in many cases. Vanishing results exist for line bundles of high degree. We look at the various degrees ν. We assume that $g(C) \geq 2$.

Case I: $\nu \geq 1$. Suppose first that L^ν is non special, i.e., $H^1(C, L^\nu) = 0$. Then $T_X^1(\nu) = H^1(C, \mathcal{D}iff_C^1 \otimes L^\nu) = H^1(C, \Theta_C \otimes L^\nu)$. In particular, if $\deg L > 4g - 4$, then $T_X^1(\nu) = 0$ [Mu3]. If L^ν is special, but $H^1(C, L^{\nu+1}) = 0$, then $T_X^1(\nu) = H^1(C, \mathcal{D}iff_C^1 \otimes L^\nu)$ still holds. If $L^\nu \neq K$, then one has $H^0(C, \Theta_C \otimes L^\nu) = 0$ and therefore the exact sequence

$$0 \longrightarrow H^1(C, L^\nu) \longrightarrow H^1(C, \mathcal{D}iff_C^1 \otimes L^\nu) \longrightarrow H^1(C, \Theta_C \otimes L^\nu) \longrightarrow 0 \,.$$

If $L^\nu = K$, the extension $0 \to K \to \mathcal{D}iff_C^1 \otimes K \to \mathcal{O}_C \to 0$ is non-trivial so the connecting homomorphism $H^0(C, \mathcal{O}_C) \to H^1(C, K)$ is an isomorphism and $T_X^1(\nu) \cong H^1(C, \mathcal{O}_C)$.

Case II: $\nu = 0$. One has $T_X^1(0) = \ker\{H^1(C, \mathcal{D}\!iff_C^1) \longrightarrow V^* \otimes H^1(C, L)\}$. The map here can also be thought of as a cup product $H^1(C, \mathcal{D}\!iff_C^1) \otimes H^0(C, L) \longrightarrow H^1(C, L)$ [AC]; for a differential operator θ and a section $s \in H^0(L)$ the cup product $\theta \cdot s = 0 \in H^1(L)$ if and only if the section s lifts to the first order deformation of $L \to C$, defined by θ. In [loc.cit.] the vector space $T_X^1(0)$ is identified as tangent space to a space of g_d^r's on a variable curve. More precisely, let $\pi: X \to S$ be a miniversal family of smooth curves, and consider the relative Picard variety $\mathrm{Pic}_{X/S}^d$ and the bundle $\mathcal{G}_d^r$ over S of g_d^r's on the fibres; there is a map $c: \mathcal{G}_d^r \to \mathrm{Pic}_{X/S}^d$ with image $\mathcal{W}_d^r$. Because by assumption the linear series L is complete, the map c is injective. Then $H^1(C, \mathcal{D}\!iff_C^1)$ is the tangent space to $\mathrm{Pic}_{X/S}^d$ in the point $L \to C$, and $T_X^1(0)$ is the tangent space to $\mathcal{G}_d^r$. The problem now is to prove that $\mathcal{G}_d^r$ is smooth of expected dimension $3g - 3 + \rho$. However, for $\rho < 0$ not much is known.

For non special L one has $\dim T_X^1(0) = 4g - 3$. In terms of deformations this means that the cone can be deformed by changing the moduli of the curve or by changing the line bundle L in Pic^d. For $L = K$ the composed map $H^1(\mathcal{O}_C) \to H^1(\mathcal{D}\!iff_C^1) \to V^* \otimes H^1(K) \cong H^1(\mathcal{O}_C)$ is an isomorphism, so $\dim T_X^1(0) = 3g - 3$ and the versal family is a family of curves in their canonical embedding.

Case III: $\nu \le -2$. In this case $T_X^1(\nu) = H^0(C, N_C \otimes L^\nu)$ because $L^{\nu+1}$ is a line bundle of negative degree. We have two vanishing results.

Lemma [Mu3]. *If $T_X^1(-1) = 0$, then $T_X^1(\nu) = 0$ for all $\nu \le -2$.*

Proof. If $H^0(N_C \otimes L^\nu) \ne 0$, then $N_C \otimes L^{-2}$ has a non-zero section s, and for all $t \in V = H^0(L)$ the tensor product $t \otimes s$ is a non-zero section of $N_C \otimes L^{-1}$; therefore $h^0(C, N_C \otimes L^{-1}) \ge \dim V$. Because $T_X^1(-1) = 0$, the map $V^* \to H^0(N_C \otimes L^{-1})$ is surjective, so all sections are of the form $t \otimes s$, therefore they are proportional, and do not generate $N_C \otimes L^{-1}$. But $N_C \otimes L^{-1}$ is generated by its sections, because $V^* \otimes \mathcal{O}_C$ is, and $V^* \otimes \mathcal{O}_C \to N_C \otimes L^{-1}$ is surjective. $\qquad\square$

Remark. The argument in the proof shows that $H^0(N_C \otimes L^\nu) = 0$, $\nu \le -2$, if $\dim T_X^1(-1) < \mathrm{rank}\, N_C - 1$.

Proposition [Wa3, 2.5]. *Let $Y \subset \mathbb{P} = \mathbb{P}(V^*)$, with $V \subset H^0(L)$, be a projective variety, defined by a system of quadratic equations f. Suppose that for every non zero quadratic equation the set of relations involving it, contains a linear relation; this property holds, if the relations are generated by linear ones. Then $H^0(Y, N_Y \otimes L^\nu) = 0$ for $\nu \le -2$.*

Proof. Let $\mathcal{I}$ be the ideal sheaf of Y. Consider the complex

$$\mathcal{O}_{\mathbb{P}}(-3)^{\oplus l} \xrightarrow{\ r\ } \mathcal{O}_{\mathbb{P}}(-2)^{\oplus k} \xrightarrow{\ f\ } \mathcal{I} \longrightarrow 0\,,$$

which is not necessarily exact at $\mathcal{O}_{\mathbb{P}}(-2)^{\oplus k}$. Dualise, twist and restrict to Y to get

$$0 \longrightarrow N_Y \otimes L^{-2} \longrightarrow \mathcal{O}_Y^{\oplus k} \xrightarrow{{}^t r} \mathcal{O}_Y(1)^{\oplus l} .$$

Let K be the kernel of the map ${}^t r \colon \mathbb{C}^k \to H^0(L)^{\oplus l}$. Here $\mathbb{C}^k$ can be identified with the dual of the vector space Q of quadratic equations. The condition that not only for these generators, but every linear combition of them is involved in a linear relation implies that $K = H^0(Y, N_Y \otimes L^\nu) = 0$.   □

By a theorem of GREEN the conditions are satisfied for an embedding of a curve with a complete linear system of degree $d \geq 2g + 3$ [Gre, Thm. 4.a.1].

Example. Let C be hyperelliptic, with involution $\pi \colon C \to \mathbb{P}^1$, and let L be very ample of degree d. Then $\phi_L(C)$ lies on a scroll $\overline{S}$, the image in $\mathbb{P}(H^0(C, L)^*)$ of $S = \mathbb{P}(\pi_* L)$, where $\pi_* L \cong \mathcal{O}(a) \oplus \mathcal{O}(b)$ with $a + b = d - (g + 1)$, and $a, b \leq d/2$. Write $e = a - b$ (we suppose $b \leq a$). Then $0 \leq e \leq g + 1$, and $S \cong \mathbb{P}(\mathcal{O} \oplus \mathcal{O}(-e))$. The Picard group of S is generated by the section E_0 with $E_0^2 = -e$, and the class f of a fibre. We have $C \sim 2E_0 + (g + 1 + e)f$ (the coefficient of f can be computed from the adjunction formula). Therefore $C^2 = 4g + 4$.

Now suppose that $d = \deg L = 2g + 2$. Then $\overline{S} = S$, except when $b = 0$, which occurs if $L = (g + 1)g_2^1$; in that case X is the cone over a rational normal curve of degree $g + 1$, and C does not pass through the vertex (because $E_0 \cdot C = 0$). We have the normal bundle exact sequence

$$0 \to N_{C/S} \longrightarrow N_C \longrightarrow N_{\overline{S}}|_C \to 0 .$$

Because the scroll $\overline{S}$ is defined by quadratic equations with linear relations, the argument of the proposition gives that $H^0(C, N_{\overline{S}} \otimes L^{-2}) = 0$. Therefore $H^0(N_C(-2)) = H^0(N_{C/S}(-2))$. Because $L \sim E_0 + af$, we have $C \cdot (C - 2L) = C \cdot (g + 1 + e - 2a)f = 0$, so $N_{C/S}(-2) \cong \mathcal{O}_C$ and $h^0(C, N_{C/S}(-2)) = 1$. For $L = (g + 1)g_2^1$ this result was obtained by DREWES [Dr].

We specialise to the case $g = 3$. The cone X over $\phi_L(C)$ has three smoothing components, if $L = 2K(= 4g_2^1)$, and two components otherwise (cf. Chap. 14, Example 5). The curve C is a complete intersection of the scroll and a quadric. If $L = 2K$, the scroll is the projective cone over a rational normal curve of degree 4, which itself has two smoothing components; the deformation to the Veronese surface occurs, if we deform C to a non hyperelliptic curve, with $L = 2K$.

We compute $T_X^1(-2)$ for $g(C) = 3$, $\deg L = 8$, C not hyperelliptic. If $L = 2K$, we find as above that $H^0(N_C(-2)) = H^0(N_{C/S}(-2)) = H^0(\mathcal{O}_C)$, where S is the Veronese surface. If $L \neq 2K$, let D be a general divisor in the linear system; it is cut out on the canonical curve C_4 in $\mathbb{P}^2$ by a cubic C_3, and the linear system is the system of cubics through the residual intersection $C_4 \cap C_3 - D$. Therefore $C \subset \mathbb{P}^5$ lies on (non unique) Del Pezzo surface of degree 5. One checks that all equations (which are quadratic) occur in linear relations, although the relations are not generated by linear ones. So $T_X^1(-2) = 0$.

Case IV: $\nu = -1$. This is the most difficult case. A more specific knowledge of the maps in the theorem is necessary. We review the relation with WAHL's Gaussian map [Wa5], see also [Te1, Dr, Wa6].

We start with the following diagram:

$$V \otimes \mathcal{O}_C$$
$$\downarrow$$
$$0 \to K \otimes L \longrightarrow \mathcal{P}_C^1(L) \longrightarrow L \to 0 .$$

The kernel of the composed map $V \otimes \mathcal{O}_C \to L$ is a vector bundle $\mathcal{M}_L$ over C, and we get a map $\mathcal{M}_L \to K \otimes L$. Let M be a second line bundle, and tensor everything with M. Define

$$\mathcal{R}(L, M) = \ker\{\mu_{L,M}: H^0(L) \otimes H^0(M) \to H^0(L \otimes M)\} .$$

Then $\mathcal{R}(L, M) = H^0(\mathcal{M}_L \otimes M)$ and we have the *Gaussian map*

$$\Phi_{L,M}: \mathcal{R}(L, M) \to H^0(K \otimes L \otimes M) .$$

This map is given explicitly by $\Phi_{L,M}(\alpha) = \sum d_L^1 l_i \otimes m_i$, where $\alpha = \sum l_i \otimes m_i \in \mathcal{R}(L, M)$ with $l_i \in H^0(L)$ and $m_i \in H^0(M)$. A more symmetric definition can be given in local coordinates; represent sections on an open set U by functions, again denoted by l_i and m_i. Then from $\sum l_i m_i = 0$ we get $\sum (l_i\, dm_i + m_i\, dl_i) = 0$, so $\Phi_{L,M}(\alpha)$ can be represented by the 1-form $1/2 \sum (l_i\, dm_i - m_i\, dl_i)$.

For curves we have that $H^1(K \otimes L \otimes M) = 0$, if $\deg L + \deg M > 0$. Denoting the map $H^0(L) \otimes H^0(M) \to H^0(\mathcal{P}_C^1(L) \otimes M)$ by $d_L^1 \otimes 1_M$, we get the exact sequence $0 \to \operatorname{coker}\Phi_{L,M} \to \operatorname{coker} d_L^1 \otimes 1_M \to \operatorname{coker}\mu_{L,M} \to 0$.

We can also apply the construction above with M instead of L. Up to a permutation of factors one has $\mu_{L,M} = \mu_{M,L}$ and $\Phi_{L,M} = -\Phi_{M,L}$. Therefore

$$\operatorname{coker} d_L^1 \otimes 1_M \cong \operatorname{coker} d_M^1 \otimes 1_L .$$

We apply this to the computation of $T_X^1(-1)$. As $(T_X^1(\nu))^* = \operatorname{coker} d_L^1 \otimes 1_{K \otimes L^{-(\nu+1)}}$, we have in particular $(T_X^1(-1))^* = \operatorname{coker} d_L^1 \otimes 1_K$. Therefore $(T_X^1(-1))^* \cong \operatorname{coker} d_K^1 \otimes 1_L$.

The multiplication maps $\mu_{L, K \otimes L^{-(\nu+1)}}$ are surjective, if the map $\phi_L: C \to \mathbb{P}(H^0(C, L)^*)$ is birational onto its image [AS, Proof of Thm 1.6]. This applies to our situation, so $(T_X^1(\nu))^* = \operatorname{coker}\Phi_{L, K \otimes L^{-(\nu+1)}}$ and in particular $(T_X^1(-1))^* = \operatorname{coker}\Phi_{L,K} = \operatorname{coker}\Phi_{K,L}$.

Now suppose that C is *not hyperelliptic*. Then K is very ample and we have the exact sequence

$$0 \longrightarrow N_K^* \otimes K \longrightarrow H^0(K) \otimes_{\mathbb{C}} \mathcal{O}_C \longrightarrow \mathcal{P}_C^1(K) \longrightarrow 0 ,$$

where N_K^* is the conormal bundle of the canonical embedding. This yields $(T_X^1(-1))^* = \ker\{H^1(N_K^* \otimes K \otimes L) \to H^0(K) \otimes H^1(L)\}$. This map is surjective, because $H^1(\mathcal{P}_C^1(K) \otimes L) = 0$. Dually

$$T_X^1(-1) = \mathrm{coker}\{H^0(K)^* \otimes H^0(K \otimes L^{-1}) \to H^0(N_K \otimes L^{-1})\} .$$

In particular:

Lemma. *For the cone X over a non hyperelliptic curve, embedded by a non special complete linear system, $T_X^1(-1) = H^0(N_K \otimes L^{-1})$.*

As MUMFORD remarks, there is an integer d_0, depending only on C, such that $H^0(N_K \otimes L^{-1}) = 0$ for $\deg L \geq d_0$. The bundle $\mathcal{E}$ of [Mu3] is $N_K \otimes K^{-1}$; in fact, MUMFORD's construction can be understood as interchanging K and L.

In general it is difficult to give sharp explicit bounds, but for low genus the result is very effective. Before we give some examples, we prove a result [Wa6, 2.2], which is based on a more general lemma of LAZARSFELD, giving sufficient conditions for surjectivity of Gaussian maps.

Lemma. *Suppose C is not hyperelliptic, trigonal or a plane quintic. If $H^0(K^2 \otimes L^{-1}) = 0$ (in particular, if $\deg L > 4g - 4$), then $\Phi_{K,L}$ is surjective.*

Proof. By Petri's Theorem the ideal I of the canonical curve C is generated by quadrics, so there is a surjection $\mathcal{O}(-2)^{\otimes a} \to I$. Dualising, restricting to C and twisting gives an injection $H^0(N_K \otimes L^{-1}) \to (H^0(K^2 \otimes L^{-1}))^{\otimes a}$. $\square$

More generally, WAHL proves that $\Phi_{L,M}$ is surjective, if L and M are line bundles of degree at least $2g + 2$ on a smooth curve of genus g and $\deg L + \deg M\ 6g + 2$.

We now look at low genera.

$g = 3$. The normal bundle is the line bundle K^4. Therefore $d_0 = 17$. For $\deg L = 16$ only $L = K^4$ has non zero $T^1(-1)$, and the cone $X(C, K^4)$ is smoothable, as hyperplane section of $X(\mathbb{P}^2, \mathcal{O}(4))$. For all L with $\deg L = 13$ the Gaussian map $\Phi_{K,L}$ is not surjective, for the general L of degree 14 it is.

$g = 4$. The normal bundle is $K^2 \oplus K^3$, so $d_0 = 19$.

More generally, for C trigonal, the canonical curve sits on a scroll S as divisor of type $3H - (g-4)f$ [Schr], and we have the normal bundle sequence $0 \to N_{C/S} \to N_C \to N_S|_C \to 0$. Here $N_{C/S} = K^3 \otimes (g_3^1)^{4-g}$, a bundle of degree $3g + 6$. Sonny TENDIAN proves that $H^1(N_S|_C \otimes L^{-1}) = 0$, if $H^0(K^2 \otimes L^{-1}) = 0$, or for $L = K^2$ [Tel]. In particular, $d_0 \leq \max(3g + 7, 4g - 4)$. We conjecture that $H^1(N_S|_C \otimes L^{-1}) = 0$ if $\deg L \geq d_1$ for some $d_1 < 3g + 3$. For $\deg L = 3g + 6$ we would then have that $T^1(-1) \neq 0$ only if L is the characteristic linear system of the family of curves of type $3H - (g-4)R$ on S, which also gives a one parameter smoothing.

$g = 5$. The general curve is a complete intersection of three quadrics, with normal bundle $3K^2$, and $d_0 = 17$ (for a trigonal ^5C we have $d_0 = 22$).

$g = 6$. Here the famous possibility of a plane quintic occurs. It gives $d_0 = 26$. The general curve lies on a (possibly singular) Del Pezzo surface S of degree 5 with normal bundle $N_{C/S} = K^2$.

TENDIAN has studied the case $L = K^2$ [Te1]. In this case Φ_{K,K^2} computes also $T^1(-2)$ for the cone over the canonical embedding. TENDIAN first proves that Φ_{K,K^2} is surjective, if the Clifford index of C is at least 3. This leaves him with hyperelliptic, trigonal and tetragonal curves, as well as smooth plane quintics and sextics. We now concentrate on the tetragonal case. Then the canonical curve C lies on a three dimensional scroll $S(e_1, e_2, e_3)$ of degree $e_1 + e_2 + e_3 = g - 3$ [Schr], as complete intersection of divisors $Y \sim 2H - b_1 R$ and $Z \sim 2H - b_2 R$ with $b_1 + b_2 = g - 5$ and $b_2 \leq b_1$. Either Y is rational, or the given g_4^1 is composed with an elliptic or hyperelliptic involution: $C \xrightarrow{2:1} E \xrightarrow{2:1} \mathbb{P}^1$, and Y is a ruled surface over E, with a rational curve of double points, which is the canonical image of E. Then $\deg Y = g - 1 + b_2$, $p_a(E) = 1/2(b_2 + 2)$, and C does not intersect the double curve. The normal bundle of C in Y is $N_{C/Y} = K^2 \otimes (g_4^1)^{-b_2}$. Equations and relations for the canonical curve can be used to compute $T^1_{X(C,K)}(-2)$ directly (see [St9]). If $b_2 > 0$, then $T^1_{X(C,K)}(-2) = 0$. In the case $b_2 = 0$ and $g > 5$ one has $\dim T^1_{X(C,K)}(-2) = 1$. The curve C lies on a a Del Pezzo surface Y or on a unique elliptic cone Y. So we have $\dim T^1_{X(C,K^2)}(-1) = 1$ and Y leads to a smoothing or to a one parameter deformation to a simple elliptic singularity of degree $g - 1$.

The examples above, where $\Phi_{K,L}$ is not surjective, have in common that N_K is unstable; the rank of N_K is $g - 2$, the degree is $2(g^2 - 1) = (g - 2)(2g + 4) + 6$, and we found surfaces Y with $\deg N_{C/Y} > 2g + 4 + \lfloor 6/g - 2 \rfloor$. This motivates:

Question. Suppose that the canonical curve $^g C$ lies on a unique surface Y, such that $d_0 := \deg N_{C/Y} > 2g + 4 + \lfloor 6/g - 2 \rfloor$. Is then $\Phi_{K,L}$ surjective for all L with $\deg L > d_0$, and if $\deg L = d_0$ for $L \neq N_{C/Y}$?

For a general curve such surfaces do not exist; already for $g = 7$ the general curve has a plane representation as $C_7(8A^2)$, so $\deg N_{C/Y} = 17$.

Question [Wa6, 2.5]. Is $\Phi_{K,L}$ surjective on a general curve of genus $g \geq 12$ for $\deg L \geq 2g - 2$? What is the best bound?

We remark that for general C with $g = 10$ or $g \geq 12$ the map $\Phi_{K,K}$ is surjective [CHM]. The answer to the first part of the question is NO [Lop]: for L of the form $K + 2P$ with P a point of the curve the map $\Phi_{K,L}$ is never surjective. Indeed, the bundle $L = K + 2P$ is not ample, but maps the curve C to a cuspidal curve Γ in $\mathbb{P}^g$. We now look at the bundle of principal parts for L. The map $H^0(L) \otimes \mathcal{O}_C \to P^1(L)$ is not surjective, but has one-dimensional cokernel.

A first bound was obtained by PAOLETTI [Pao]. The problem is of course to find a general curve; the easiest to handle are singular curves. As mentioned in Chap. 12, I constructed for all $g \geq 5$ a g-cuspidal canonically embedded curve Γ with a Weil divisor D of degree $g + 5$ such that $h^0(N_\Gamma(-D)) = 6$ [St1]. It follows that $h^0(N_K \otimes L^{-1}) \leq 6$ for a general curve C with $g \geq 5$ and a general L with $\deg L = g + 5$. So $H^0(N_K \otimes L^{-1}) = 0$ for a general L

of degree at least $g + 11$; presumably a general L of degree $g + 6$ will do for large g.

Proposition. *For a general curve of genus $g \geq 3$ the map $\Phi_{K,L}$ is surjective for all L with $\deg L \geq 2g + 11$.*

Proof. Fix a line bundle L_0 of degree $g+11$ with $H^0(N_K \otimes L_0^{-1}) = 0$. Every line bundle L with $\deg L \geq 2g + 11$ can be written as $L = L_0(D)$ with D an effective divisor of degree $\deg L - g - 11 \geq g$. This implies that $H^0(N_K \otimes L^{-1})$ is a subspace of $H^0(N_K \otimes L_0^{-1})$. For $g = 3, 4$ see above. $\square$

The best known bound is $2g + 9$ for odd g and $2g + 8$ for even g, $g \geq 22$ [Par].

The hyperelliptic case. The formula

$$\left(T_X^1(-1)\right)^* = \operatorname{coker} d_L^1 \otimes 1_K \cong \operatorname{coker} d_K^1 \otimes 1_L$$

still holds. However the map $d_K^1 \colon H^0(K) \otimes \mathcal{O}_C \to \mathcal{P}^1(K)$ is only generically surjective, with cokernel equal to $\operatorname{coker}\{\phi_K^* \Omega_{\mathbb{P}}^1 \otimes K \to K^2\} \cong \Omega_\phi \otimes K$, where Ω_ϕ are the relative differentials, so the cokernel is given by the Jacobian ideal [Pi]. Therefore, if B denotes the set of Weierstraß points, we have an exact sequence

$$H^0(K) \otimes \mathcal{O}_C \xrightarrow{d_K^1} \mathcal{P}^1(K) \longrightarrow \bigoplus_{p \in B} \mathbb{C}_p \to 0 \,.$$

Let $\mathcal{P} = \operatorname{Im} d_K^1$; this is a rank 2 vector bundle.

Lemma. *Let the rational normal curve $R = \phi_K(C)$ be the canonical image of C, and write $\phi \colon C \to R$; let $H = \mathcal{O}_R(1) \cong \mathcal{O}_{\mathbb{P}^1}(g - 1)$. Let N_R be the normal bundle of R in $\mathbb{P}^{g-1}$. The following sequence is exact:*

$$0 \to \phi^*(N_R^* \otimes H) \longrightarrow H^0(K) \otimes \mathcal{O}_C \xrightarrow{d_K^1} \mathcal{P} \to 0 \,.$$

Proof. Consider the commutative diagram

$$
\begin{array}{ccccccccc}
0 & \to & \phi^*(N_R^* \otimes H) & \longrightarrow & H^0(K) \otimes \phi^*(\mathcal{O}_R) & \xrightarrow{\phi^* \circ d_H^1} & \mathcal{P}_R^1(K) & \to & 0 \\
& & \downarrow & & \downarrow{\scriptstyle \cong} & & \downarrow & & \\
0 & \to & \mathcal{E} & \longrightarrow & H^0(K) \otimes \mathcal{O}_C & \xrightarrow{d_K^1} & \mathcal{P}_C^1(K) & &
\end{array}
$$

Because $\phi^* \circ d_H^1 = d_K^1 \circ \phi^*$ [EGA, iV.16.4.3.4] the image of the right-hand vertical map is $\mathcal{P}$, and therefore $\mathcal{E} = \phi^*(N_R^* \otimes H)$. $\square$

The normal bundle of R splits as a direct sum of $g - 2$ bundles of degree $g+1$, and therefore $\mathcal{E} = \oplus_{g-2}(g_2^1)^{-2}$. Using the fact that $H^1(\mathcal{P}_C^1(K) \otimes L) = 0$ for $\deg L > 0$, we obtain the following result; for $\deg L > 2g + 1$ cf. [Wa4, 7.11.1], [Dr].

Proposition. *Let C be a hyperelliptic curve of genus g, and L a line bundle with $\deg L > 0$. Then $\dim \operatorname{coker} d_K^1 \otimes 1_L = 2g + 2 + (g-2)h^1(L \otimes (g_2^1)^{-2}) - gh^1(L)$. In particular, if $\deg L > 2g + 2$, then $\dim T^1(-1) = 2g + 2$.*

We remark that if the line bundle L defines a birational map of C, then $\pi_* L \cong \mathcal{O}_{\mathbb{P}^1}(a) \oplus \mathcal{O}_{\mathbb{P}^1}(b)$ with $a > b \geq 0$, and $h^1(L \otimes (g_2^1)^{-2}) \leq 1$.

T^2 and obstructions. For a reduced singularity X we found in Chapter 10 the following description of T_X^2:

$$0 \to N_X \longrightarrow \operatorname{Hom}\left((\mathcal{O}_X)^{\oplus k}, \mathcal{O}_X\right) \longrightarrow \operatorname{Hom}(R_X, \mathcal{O}_X) \longrightarrow T_X^2 \to 0 .$$

If the sheaf $\mathcal{T}^2$ has support contained in a subspace Z with $\operatorname{dp}_Z X \geq 2$, then

$$\begin{aligned}
T_X^2 &= \operatorname{coker}\left\{H^0(U, \mathcal{O}_n^{\oplus k}) \longrightarrow H^0(U, R_X^*)\right\} \\
&= \ker\left\{H^1(U, N_X) \longrightarrow H^1\left(U, (\mathcal{O}_n)^{\oplus k}\right)\right\} ,
\end{aligned}$$

where $U = X \setminus Z$.

We specialise to the case of cones as before, so let X be the cone over the smooth curve C, embedded with the line bundle L. Let the ideal of C in $\mathbb{P}(V^*)$, $V = H^0(C, L)$, be generated by k equations of degree $d_1, \ldots, d_k$. Then the graded parts of T_X^2 are given by the exact sequence

$$0 \to T_X^2(\nu) \longrightarrow H^1(N_C(\nu)) \longrightarrow H^1(\mathcal{O}_C(d_1 + \nu)) \oplus \cdots \oplus H^1(\mathcal{O}_C(d_k + \nu)) .$$

As we have seen, the group $H^1(N_C(\nu))$ occurs in the following exact sequence:

$$0 \to T_X^1(\nu) \longrightarrow H^1\left(\mathcal{D}\mathit{iff}_C^1(\nu)\right) \longrightarrow V^* \otimes H^1(L^{\nu+1}) \longrightarrow H^1(N_C(\nu)) \to 0 .$$

This gives the following formula for the dimension of $T_X^2(\nu)$:

$$\dim \ker\{V^* \otimes H^1(L^{\nu+1}) \to \oplus_i H^1(L^{d_i+\nu})\} - h^1\left(\mathcal{D}\mathit{iff}_C^1(\nu)\right) + \dim T_X^1(\nu) .$$

If $H^1(C, L) = 0$, then $H^1(C, N_C(\nu)) = 0$ for $\nu \geq 0$, and therefore also $T_X^2(\nu) = 0$. If C is defined by quadratic equations, and L is not special, then $T_X^2(-1) = H^1(C, N_C(-1))$ and therefore

$$\dim T_X^2(-1) = (g-2)h^0(L) - 6(g-1) + \dim T_X^1(-1) ,$$

because $h^1\left(\mathcal{D}\mathit{iff}_C^1(-1)\right) = 2 \deg L + 4g - 4$.

Example. The above computation gives new examples of singularities for which the obstruction map is not surjective, cf. [Te1, 2.4.1]. In particular, if $g(C) = 3$ and L is general of degree $d \geq 14$, then $\dim T_X^1(-1) = 0$, and X has only conical deformations, so the base space is smooth, whereas $\dim T_X^2(-1) = d - 14$. For $g = 4$ and general L with $9 \leq d = \deg L < 15$ we have $\dim T_X^1(-1) = 15 - d$, $\dim T_X^1 = 28 - d$, which is also the dimension of a smoothing component [Te2, 6.2]. Therefore the base space is smooth, but $\dim T_X^2(-1) = d - 9$. Finally, for a general $X(^5C, L_d)$ with $d > 12$ there are only conical deformations, and $\dim T_X^2(-1) = 3(d - 12)$.

For $T_X^2(\nu)$ with $\nu < -2$ we have the following general vanishing result:

Lemma [Wa6, Cor. 2.10]. *If $Y \subset \mathbb{P} = \mathbb{P}^n$ is a smooth projectively normal subvariety, defined by quadratic equations, with a resolution*

$$\mathcal{O}_{\mathbb{P}}(-4)^{\oplus m} \longrightarrow \mathcal{O}_{\mathbb{P}}(-3)^{\oplus l} \longrightarrow \mathcal{O}_{\mathbb{P}}(-2)^{\oplus k} \longrightarrow \mathcal{O}_{\mathbb{P}} \longrightarrow \mathcal{O}_X \longrightarrow 0 \ ,$$

then the cone X over Y satisfies $T_X^2(\nu) = 0$ for $\nu < -2$.

By [Gre, Thm. 4.a.1] the conditions are satisfied for a curve, embedded with a complete linear system of degree $d \geq 2g + 4$.

Smoothings. Let $\overline{X} \subset \mathbb{P}^{N+1}$ be the projective cone over the curve $C \subset \mathbb{P}^N$, and let $S \subset \mathbb{P}^{N+1}$ be a surface with C as hyperplane section. Then there exists a one parameter deformation of $\overline{X}$ with S as general fibre, by 'sweeping out the cone' over S. For the affine cone X over C we have a deformation with Milnor fibre $S - C$. The versal base in negative degree is a fine moduli space for so called R-polarised schemes [Loo], a notion defined in general for quasi-homogeneous spaces; in our situation this are spaces S with C as hyperplane section. Basically one considers the coordinate t, which defines the hyperplane section, as deformation parameter.

Let S be given by equations $F_i(\boldsymbol{x}, t) = f_i(\boldsymbol{x}) + tf_i^{(1)}(\boldsymbol{x}) + \ldots + t^{d_i} f_i^{(d_i)}$, in homogeneous coordinates $(\boldsymbol{x}, t)$, where $\deg f_i^{(j)} = d_i - j$. The base space of X has a $\mathbb{C}^*$-action. For simplicity we assume that the only occurring negative degree is -1. The infinitesimal deformation of X, induced by S, is given by $f_i \mapsto \frac{\partial}{\partial t} F_i(\boldsymbol{x}, t)|_{t=0} = f_i^{(1)}$. This formula can be interpreted as defining a section of $H^0(C, N_C(-1))$.

Let S be embedded by the line bundle L, and let $\mathbb{P} = \mathbb{P}(H^0(S, L)^*)$. Let $C = H \cap S$, with H a hyperplane in $\mathbb{P}$, and let $i : C \to S$ be the inclusion. Then $N_{C/S} = L_{|C}$. In the exact sequence

$$0 \longrightarrow \mathcal{O}_C \cong N_{C/S}(-1) \longrightarrow N_{C/\mathbb{P}}(-1) \longrightarrow i^* N_{S/\mathbb{P}}(-1) \longrightarrow 0$$

the infinitesimal deformation is the image of $1 \in \mathcal{O}_C$. The curve C is minimally embedded in the hyperplane H, so we really want a section of $N_{C/H}(-1)$. The exact sequence

$$0 \longrightarrow N_{C/H}(-1) \longrightarrow N_{C/\mathbb{P}}(-1) \longrightarrow i^* N_{H/\mathbb{P}}(-1) \longrightarrow 0$$

splits: $N_{H/\mathbb{P}}(-1)$ is generated by the global section, which sends the equation t to 1, and this section can be mapped to the section $F \mapsto \frac{\partial}{\partial t} F|_{t=0}$ of $N_{C/\mathbb{P}}(-1)$.

The above applies in all dimensions, but for curves we have a third description, which uses $(T^1)^*$. We have seen that $T_X^1(-1)^* = \operatorname{coker}\{V \otimes H^0(K) \to H^0(\mathcal{P}_C^1(L) \otimes K)\}$. Consider the following diagram of exact sequences:

$$
\begin{array}{ccccccccc}
& & & & 0 & & 0 & & \\
& & & & \downarrow & & \downarrow & & \\
0 & \longrightarrow & \mathcal{O}_C & \longrightarrow & \Omega^1_S \otimes \mathcal{O}_C(1) & \longrightarrow & \Omega^1_C(1) & \longrightarrow & 0 \\
& & \| & & \downarrow & & \downarrow & & \\
0 & \longrightarrow & \mathcal{O}_C & \longrightarrow & \mathcal{P}^1_S(L) \otimes \mathcal{O}_C & \longrightarrow & \mathcal{P}^1_C(L) & \longrightarrow & 0 \\
& & & & \downarrow & & \downarrow & & \\
& & & & L & = & L & & \\
& & & & \downarrow & & \downarrow & & \\
& & & & 0 & & 0 & &
\end{array}
$$

Proposition. Let $\xi \in H^0(C, N_C(-1))$ be the infinitesimal deformation of the cone over C induced by S. The map $\cup\,\xi : T^1(-1)^* \to \mathbb{C}$ is the connecting homomorphism $H^0(\mathcal{P}^1_C(L) \otimes K) \to H^1(K)$, obtained from the above sequence by tensoring with $K = \Omega^1_C$; alternatively one may consider the homomorphism $H^0(K^2 \otimes L) \to H^1(K)$.

Proof. We first describe $H^0(\mathcal{P}^1(L) \otimes K)$. The sheaf $\mathcal{P}^1(L)$ is generated by global sections; if we take a local coordinate u on C and write a Newton dot for the derivative with respect to u, then the map $V \otimes \mathcal{O}_C \to \mathcal{P}^1(L)$ is given, in accordance with our earlier notations, by $dz_i \mapsto z_i(u) + \dot{z}_i(u)du$. Let φ be a global section of $\mathcal{P}^1(L) \otimes K$. On $U_j = \{z_j \neq 0\}$ the section φ can be represented as $\sum \omega^i_j dz_i$, and the representations for different j are connected by the conditions $\sum \omega^i_j z_i = \sum \omega^i_k z_i$ and $\sum \omega^i_j \dot{z}_i = \sum \omega^i_k \dot{z}_i$.

This allows us to compute $\varphi \cup \xi$: the cochain $\sum(\omega^i_k - \omega^i_j)dz_i$ represents an element of $H^1(N^*_C(1) \otimes \Omega^1_C) = H^0(N_C(-1))^*$. To express ξ in terms of the $\frac{\partial}{\partial z_i}$, we take local coordinates (u, τ) on S with τ vanishing on C, and we denote differentiation w.r.t. τ by $'$. For every equation F of S we have $\partial F/\partial \tau = 0$, so

$$
t'(u,0)\frac{\partial F}{\partial t} + \sum z'_i(u,0)\frac{\partial F}{\partial z_i} = 0 \,,
$$

and therefore $\xi = -\sum(z'_i/t')\frac{\partial}{\partial z_i}$. Now $\varphi \cup \xi = \sum(\omega^i_j - \omega^i_k)(z'_i/t')$.

To compute the connecting homomorphism δ in the exact sequence

$$
0 \to \Omega^1_C \to \mathcal{P}^1_S(L) \otimes \Omega^1_C \to \mathcal{P}^1_C(L) \otimes \Omega^1_C \to 0 \,,
$$

we lift φ on U_j to $\mathcal{P}^1(L) \otimes K$. On $U_j = \{z_j \neq 0\}$ we write φ as $\sum \omega^i_j dz_i + \omega_j dt$, and the cochain conditions on S are: $\sum \omega^i_j z_i + \omega_j t = \sum \omega^i_k z_i + \omega_k t$, $\sum \omega^i_j \dot{z}_i + \omega_j \dot{t} = \sum \omega^i_k \dot{z}_i + \omega_k \dot{t}$ and $\sum \omega^i_j z'_i + \omega_j t' = \sum \omega^i_k z'_i + \omega_k t'$; for $\tau = 0$ we have $t = \dot{t} = 0$, so the first two conditions are the same as before, and we have $(\omega_k - \omega_j)t' = \sum(\omega^i_j - \omega^i_k)z'_i$. The cochain $\omega_k - \omega_j$ represents $\delta(\varphi) \in H^1(K)$, and therefore $\delta(\varphi) = \varphi \cup \xi$. $\qquad\square$

16 The versal deformation of hyperelliptic cones

We consider the cone $X = X(C, L)$ over a hyperelliptic curve ${}^g C$, embedded with a complete linear system L of degree at least $2g + 3$. Then, as we have seen in the previous Chapter, T_X^1 is concentrated in degrees 1, 0 and -1, and $T_X^1(1) = 0$ if $\deg L > 4g - 4$. We restrict ourselves to the part of the versal deformation in negative degree, because otherwise also non hyperelliptic curves come in.

For the computation of T^1 efficient methods exist, which avoid the explicit use of equations and relations. For the versal deformation there seems to be no alternative. Actually, as Frank SCHREYER repeatedly pointed out to me, the equations for $\phi_L(C)$ are rather simple in the hyperelliptic case: C is a divisor on a two-dimensional scroll, so besides the determinantal of the scroll we have 'essentially' one equation.

Equations for X. Let L be any line bundle of degree $d \geq 2g + 3$. Denote the involution by $\pi : C \to \mathbb{P}^1$. Then $\phi_L(C)$ lies on the scroll $S = \mathbb{P}_{\mathbb{P}^1}(\pi_* L)$, where $\pi_* L \cong \mathcal{O}(a) \oplus \mathcal{O}(b)$ with $a + b = d - (g + 1)$, and $a, b \leq d/2$; in particular, if $L = kg_2^1$, then $a = k$ and $b = k - (g + 1)$. Suppose $b \leq a$, write $e = a - b$, so $0 \leq e \leq g + 1$, and $S \cong \mathbb{P}(\mathcal{O} \oplus \mathcal{O}(-e))$. In $\operatorname{Pic} S = \mathbb{Z} E_0 \oplus \mathbb{Z} f$, where E_0 is the section with $E_0^2 = -e$ and f is the class of a fibre, we have $C \sim 2E_0 + (g + 1 + e)f$.

Let C be the curve $y^2 - \sum_{i=0}^{2g+2} a_i x^i = 0$. We can make this equation quasi-homogeneous by introducing homogeneous coordinates $(x, \hat{x})$ on $\mathbb{P}^1$:

$$C = \{y^2 - \sum_{i=0}^{2g+2} a_i \hat{x}^{2g+2-i} x^i = 0\} .$$

We write $L = kg_2^1 + D$ with $\deg D = g + 1 - e$, and k maximal such that D is effective. The divisor D can be described by two polynomials $U(x)$ and $V(x)$, together with $F(x) = \sum_{i=0}^{2g+2} a_i x^i$. Suppose $D = P_1 + \cdots + P_{g+1-e}$ with all P_i distinct and let (x_i, y_i) be the coordinates of P_i. Define $U(x) = \prod_i (x - x_i)$, and take $V(x)$ the unique polynomial of degree $\leq g - e$ with $V(x_i) = y_i$. The ideal $(U, y - V)$ defines D; this is indeed a subvariety of $y^2 - F$, and one has $F - V^2 = UW$ for some polynomial W.

The function $(y + V)/U$ defines a section of $H^0(C, L)$; a basis of this vector space can be represented in inhomogeneous coordinates by the polynomial forms

$$z_i = Ux^i, \qquad i = 0, \ldots, k \,,$$
$$w_i = (y + V)x^i, \qquad i = 0, \ldots, k - e \,.$$

The equations for X are those for the scroll, and $2k - 2g - 1$ further equations ϕ_m, obtained by *rolling factors* (Chap. 12, Example 4). Modulo the equation $y^2 - F$ we have the relation $(y+V)^2 - 2(y+V)V - UW = 0$; we multiply this equation with U and suitable powers of $\hat{x}$ and x. Write $U(x) = \sum_{i=0}^{g+1-e} U_i x^i$, $V(x) = \sum_{i=0}^{g-e} V_i x^i$ and $W(x) = \sum_{i=0}^{g+1+e} W_i x^i$. To avoid making a specific choice for the equations ϕ_m we denote by $(w^2)_n$ any product $w_p w_q$ with $p + q = n$.

Then we have the following equations:

$$\mathrm{Rank}\begin{pmatrix} z_0 & z_1 & \cdots & z_{k-1} & w_0 & \cdots & w_{k-e-1} \\ z_1 & z_2 & \cdots & z_k & w_1 & \cdots & w_{k-e} \end{pmatrix} \le 1 \,,$$

$$\phi_m = \sum_{i=0}^{g+1-e} U_i(w^2)_{m+i} - 2\sum_{i=0}^{g-e} V_i(wz)_{m+i} - \sum_{i=0}^{g+1+e} W_i(z^2)_{m+i} \,,$$

$$m = 0, \ldots, d - 2g - 2 \,.$$

Infinitesimal deformations. For the cone X over a hyperelliptic curve of degree at least $2g + 3$ the dimension of $T_X^1(-1)$ is $2g + 2$ (p. 134). All deformations are of rolling factors type. To show this we study the normal bundle of C.

Let $d = \deg L \ge 2g + 3$. For the scroll $S = \mathbb{P}(\pi_* L)$, we have $\mathcal{O}_S(1) \sim E_0 + af$, so the divisor of the line bundle $N_{C/S}(-1)$ is $C \cdot (L + (2g+2-d)f)$. From the exact sequence

$$0 \to H^0(N_{C/S}(-1)) \longrightarrow H^0(N_C(-1)) \longrightarrow H^0(N_S|_C(-1))$$
$$\longrightarrow H^1(N_{C/S}(-1))$$

we obtain that $h^0(C, N_S|_C(-1)) \le h^0(C, N_C(-1)) - \chi(N_{C/S}(-1)) = d+g+3 - (3g + 5 - d) = 2d - 2g - 2$. The dimension of $H^0(C, N_C(-1))$ can be computed with the sequence

$$0 \longrightarrow V^* \otimes H^0(C, \mathcal{O}_C) \longrightarrow H^0(C, N_C(-1)) \longrightarrow T_X^1(-1) \longrightarrow 0 \,.$$

Denote the projection of the scroll by $\pi \colon S \to \mathbb{P}^1$ (it induces the hyperelliptic involution π on C). A simple computation shows that the restriction $N_S|_f$ of the normal bundle N_S to a fibre f of π splits as $\mathcal{O}_{\mathbb{P}^1}(2) \oplus \mathcal{O}_{\mathbb{P}^1}(1) \oplus \ldots \oplus \mathcal{O}_{\mathbb{P}^1}(1)$. Therefore $\pi_* N_S(-C - L) = 0$, and the the restriction map $H^0(S, N_S(-1)) \to H^0(C, N_S|_C(-1))$ is injective. Deforming the matrix of the scroll leads to deformations in degree -1, so $h^0(S, N_S(-1)) \ge 2d - 2g - 2$. We obtain that $h^0(C, N_S|_C(-1)) = 2d - 2g - 2$.

Every infinitesimal deformation of weight -1 of X induces a deformation of the scroll, but not every deformation of the scroll extends to a deformation of X. One obtains linear equations, whose coefficients come from the polynomials U, V and W. The general formula is given in [St9, Prop. 2.11]. To give an impression we describe the simplest case $L = kg_2^1$. Then $N_{C/S}(-1) = (2g + 2 - k)g_2^1$. Perturbations of the w-variables in the matrix never lift. So we perturb only the z-variables:

$$\text{Rank}\begin{pmatrix} z_0 & \cdots & z_{k-2} & z_{k-1} & w_0 & \cdots & w_{k-e-1} \\ z_1 + \zeta_1 & \cdots & z_{k-1} + \zeta_{k-1} & z_k & w_1 & \cdots & w_{k-e} \end{pmatrix} \leq 1\,.$$

If $g+1 < k \leq 2g+2$, all $k-1$ deformations $z_i + \zeta_i$ lift and we can perturb the $2k - 2g - 1$ additional equations with $2g + 3 - k$ different linear terms in rolling factors format:

$$\phi_m(z, \zeta) + \sum_{i=0}^{2g+2-k} \rho_i z_{m+i}\,, \qquad m = 0, \ldots, 2k - 2g - 2\,.$$

In the perturbations all variables $z_0, \ldots, z_k$ occur.

But if $k > 2g + 2$ these perturbations are not possible. Now there are more than $2g + 2$ variables ζ_i. Only the $(2g + 2)$-dimensional subspace, given by the $k - 2g - 2$ equations $\sum a_i \zeta_{i+j} = 0$, lift. In this case the polynomial W is equal to F, so its coefficients are the a_i.

Obstructions. We are interested in determining the versal deformation in negative degree. The general discussion of 'rolling factors' deformations of a divisor of type $2H - bf$ on a scroll (Chap. 12) shows that the base space is defined by $d - 2g - 3$ equations. As the line bundle L is non special, we can apply the formulas and vanishing results of Chap. 15 to $T_X^2(\nu)$ for $\nu \neq -2$. We obtain

Proposition. *Let $X = X(C, L)$ be the cone over a hyperelliptic curve of genus g, embedded with a complete linear system L of degree d, with $d > 2g + 3$. Then $\dim T_X^2(-2) \geq d - 2g - 3$, $\dim T_X^2(-1) = (g - 2)(d - g - 3)$, and $\dim T_X^2(\nu) = 0$ for $\nu \neq -1, -2$.*

Remark. The general hyperplane section Y of X is the cone over d distinct points on a rational normal curve of degree $d - g - 1$. One can compute that $\dim T_Y^1 = d + (g - 1)(d - g - 1)$. By the MAIN LEMMA of [BC] the number of generators T_X^2 equals $\dim T_Y^1 - e$, where $e = \mu + t - 1$ [Greu, 2.5.(3)] is the dimension of a smoothing component. In our case $t = g = \delta(Y) - d + 1$, so $e = 3g + d - 2$, and the number of generators of T_X^2 is $(g - 1)(d - g - 4) - 1$. If $g = 2$, then T_X^2 is concentrated in degree -2, so $\dim T_X^2 = d - 7$. If for general g the module T_X^2 is annihilated by the maximal ideal, then we can conclude also in that case that $\dim T_X^2(-2) = d - 2g - 3$. For $L = kg_2^1$ I verified this formula by explicitly computing the dimension of $\ker\{V^* \otimes H^1(L^{-1}) \to \oplus_i H^1(\mathcal{O})\}$. For our present purposes it suffices to know that $d - 2g - 3$ is the number of equations for the 'rolling factors' deformations.

Smoothing components. The cone over a hyperelliptic curve of degree $d \leq 4g + 4$ is smoothable. More precisely, we have:

Proposition [Te3]. *Let ^gC be a hyperelliptic curve, embedded with a complete linear system L of degree $7g/3 + 1 \leq d \leq 4g + 4$. Then C is a hyperplane section of a projectively normal surface $\overline{S} = \phi_{|C|}(S)$, where S can be obtained from a rational ruled surface by blowing up $4g + 4 - d$ points. The dimension of the smoothing component is $7g + 4 - d + h^0(C, K^2L^{-1})$. If $d > \max\{4g - 4, 3g + 6\}$ for $g \neq 6$, and $d > 25$ for $g = 6$, then every smoothing component is of this form, and has dimension $7g + 4 - d$.*

In particular, for $d = 4g + 4$ the surface S is ruled, and the dimension of the smoothing component is $3g = (2g - 1) + g + 1$. The dimension of the hyperelliptic locus is $2g - 1$, and g is the dimension of Pic^d. So for general (C, L) each smoothing component determines a unique surface $\overline{S}$ with C as hyperplane section. Furthermore $\dim T^1_X(-1) = 2g + 2$, and the base space in negative degree is given by $2g + 1$ equations. We would like to conclude that the equations define a complete intersection, but this is not possible from these numerical data alone.

The existence of smoothings can be shown in the following way: given a line bundle M of degree $4g + 4$ on C (which is in general not the linear system L for which we want a smoothing), we consider the scroll S of type (a, b), on which $\phi_M(C)$ lies. On S the hyperplane class is $M = E_0 + af$, and the curve C is a divisor of type $2E_0 + (g + 1 + e)f = 2M - (2g + 2)f$. So for every M with $M^2 \cong L \otimes (g_2^1)^{2g+2}$ the normal bundle $N_{C/S}$ is isomorphic to our fixed bundle L. The number of solutions to this equation is the order of the group $J_2(C)$ of 2-torsion points on $\mathrm{Jac}(C)$, which is 2^{2g}.

For all surfaces S obtained by this construction we have $e \equiv g + 1$ (mod 2), because $e = a - b$, and $a + b = 3g + 3$. This is not surprising, because for a fixed curve and variable L the surfaces $\mathbb{P}(\pi_*L)$ form a continuous family, and the parity of e is preserved under deformations.

To construct ruled surfaces $C \subset S \cong F_e$ with $N_{C/S} = L$ and $g + 1 + e$ odd we apply elementary transformations. The elementary transformation elm_Q of S in the point $Q \in S$ is obtained by first blowing up the point Q and then contracting the strict transform of the fibre through Q [Ha, V.5.1.7]. We have the following simple, but important observation.

Lemma. *Let C' be the strict transform of C on $S' = \mathrm{elm}_P(S)$, where P is a Weierstraß point. Then the normal bundle $N_{C'/S'}$ is equal to the normal bundle $N_{C/S}$.*

Proposition. *Let B be the set of Weierstraß points on C. Let elm_B be the composition of the elementary transformations in all points $P \in B$. Then $\mathrm{elm}_B(S)$ is isomorphic to S, under an isomorphism I, which leaves C pointwise fixed; on the general fibre f of $S \to \mathbb{P}^1$, I restricts to the unique involution with $C \cap f$ as fixed points. The isomorphism I maps the linear system $|C|$ on S to the linear system $|C'|$ on S'.*

Proof. The rational map elm_B can be factorised as $S \xleftarrow{\sigma} \tilde{S} \xrightarrow{\sigma'} S'$, where the maps σ and σ' are the blow-ups in the points of B with exceptional curves E_i and E_i'. The involution on the general fibre f of $\tilde{S} \to \mathbb{P}^1$ with $f \cap C$ as fixed points extends to an involution $\tilde{I}$ on $\tilde{S}$, which interchanges E_i and E_i'. This map descends to the required isomorphism. The inverse image on $\tilde{S}$ of the linear system $|C|$ is $|\overline{C} + \sum E_i|$, where $\overline{C}$ is the strict transform of C. The involution $\tilde{I}$ transforms this system into $|\overline{C} + \sum E_i'| = (\sigma')^*|C'|$.   □

Proposition. *Let $C \subset S \cong F_e$ be a curve of type $2E_0 + (g + 1 + e)f$ with normal bundle $N_{C/S} = N$. Denote for a subset $T \subset B$ by elm_T the composition of the elm_{P_i}, $P_i \in T$. Two surfaces $\text{elm}_{T_1}(S)$ and $\text{elm}_{T_2}(S)$ induce the same deformation of the cone $X(C, N)$ if and only if $T_1 = T_2$ or $T_1 = B \setminus T_2$.*

Proof. According to the Proposition on p. 136 we have to compute the connecting homomorphism from the exact sequence

$$0 \to \Omega_C^1 \to \Omega_S^1 \otimes \Omega_C^1(1) \to (\Omega_C^1)^{\otimes 2}(1) \to 0,$$

or, what amounts to the same, the hyperplane in $H^0(C, K^2(1))$, which is the image of $H^0(C, \Omega_S^1 \otimes \Omega_C^1(1))$. We have on S the sequence

$$0 \to \pi^* \Omega_{\mathbb{P}^1}^1 \to \Omega_S^1 \to \Omega_{S/\mathbb{P}^1}^1 \to 0$$

and on C

$$0 \to \pi^* \Omega_{\mathbb{P}^1}^1 \to \Omega_C^1 \to \Omega_{C/\mathbb{P}^1}^1 \to 0,$$

where $\Omega_{C/\mathbb{P}^1}^1 = T_X^1(-1)^* \cong \oplus_{P \in B} \mathbb{C}_P$. The map $\Omega_S^1 \otimes \Omega_C^1(1) \to (\Omega_C^1)^{\otimes 2}(1)$ is an isomorphism on the subspace $\pi^* \Omega_{\mathbb{P}^1}^1 \otimes \Omega_C^1(1)$. Because $K_S = -2E_0 - (e + 2)f$, we have $\Omega_{S/\mathbb{P}^1}^1 = \mathcal{O}_S(-2E_0 - ef)$, and $\Omega_{S/\mathbb{P}^1}^1(C + (g - 1)f) = \mathcal{O}_S(2gf)$. We identify $T_X^1(-1)^*$ with $H^0(C, (\Omega_C^1/\pi^* \Omega_{\mathbb{P}^1}^1) \otimes \Omega_C^1(1))$ by taking in each Weierstraß point P_i a generator of $K^2(1)/\mathfrak{m}_{P_i} K^2(1)$; the map from $H^0(C, \Omega_{S/\mathbb{P}^1}^1 \otimes \Omega_C^1(1))$ consists in taking the coefficients in the points P_i. The sections of $(g_2^1)^{\otimes 2g}$ are $1, \ldots, x^{2g}$, which come from sections on S, and $y, \ldots, yx^{g-1}$, which vanish in the Weierstraß points. Therefore, if we take coordinates s_i on $T_X^1(-1)^*$, and if x-coordinate of P_i is a_i, then the image of $H^0(C, \Omega_{S/\mathbb{P}^1}^1 \otimes \Omega_C^1(1))$ is given by the determinant

$$D(s_1, \ldots, s_{2g+2}) = \begin{vmatrix} 1 & a_1 & \cdots & a_1^{2g} & s_1 \\ \vdots & \vdots & \ddots & \vdots & \vdots \\ 1 & a_{2g+2} & \cdots & a_{2g+2}^{2g} & s_{2g+2} \end{vmatrix}.$$

Now we consider the surface $S' = \text{elm}_T(S)$ for some $T \subset B$. Again we can identify the sections of $H^0(C, \Omega_{S'/\mathbb{P}^1}^1 \otimes \Omega_C^1(1))$, coming from S', with the polynomials $1, \ldots, x^{2g}$, but we have to express these in the same basis of

$T_X^1(-1)^*$ used for S. Consider local coordinates (x, y) in a neighbourhood of a point $P \in T$, such that ruling is given by $\pi(x, y) = x$, and C is defined by $y^2 = x$. We blow S up in P; the strict transform of C passes through the origin of the (η, y) coordinate patch, where $(x, y) = (\eta y, y)$. Now blow down the y-axis: we have coordinates $(\xi, \eta) = (\eta y, \eta)$, so $x = \xi$, $y = \xi/\eta$. Therefore the local generator dy of $\Omega_{C,P}^1$ is transformed into $d\eta$: we have $dy = d\xi/\eta - \xi d\eta/\eta^2$; however, considered as section of $\Omega_S^1/\pi^*\Omega_{\mathbb{P}^1}^1$, the formula makes sense for $\eta \neq 0$ and is on C the same as $-\xi d\eta/\eta^2 = -d\eta$. This computation shows that S' yields the hyperplane

$$D((-1)^{\chi_T(P_1)} s_1, \ldots, (-1)^{\chi_T(P_{2g+2})} s_{2g+2}) = 0 \, ,$$

where χ_T is the characteristic function of T, i.e $\chi_T(P_i) = 1$ if and only if $P_i \in T$.

To finish the proof we remark that the coefficient of s_i in the equation D is a Vandermonde determinant, and therefore non zero.   □

Theorem. *Let X be the cone over a hyperelliptic curve C of genus g, embedded with a complete linear system L of degree $4g + 4$. Suppose $L \neq 4K$, if $g = 3$. Then X has 2^{2g+1} smoothing components.*

Proof. The number of subsets T of B modulo the equivalence relation $T \sim B\backslash T$ is 2^{2g+1}. So the previous Proposition gives this number of one parameter smoothings of X. After a rescaling these define in $\mathbb{P}(T_X^1(-1))$ the points $(\pm 1 : \ldots : \pm 1)$, and this is a complete intersection. In coordinates s_i on $\mathbb{P}(T_X^1(-1))$ the ideal of these points is generated by $s_i^2 - s_j^2$. The $2g + 1$ quadratic equations for the base in negative degree are contained in this ideal; they generate it if and only if they are linearly independent. Because the dimension of each smoothing component is $3g$, the equations are independent for generic (C, L). Suppose that for some special (C, L) the equations are dependent. Then the base space of $X(C, L)$ is not smooth along at least one of the parameter lines of the 2^{2g+1} smoothings we just constructed. But the only singularities in the fibres are cones over a rational normal curve of degree $g + 1$, which have a smooth reduced base space, if $g \neq 3$. Therefore the equations are always independent.

For $g = 3$ we still have the same description of the base in negative degree, but in case $L = 4K$ the cone over the rational normal curve of degree 4 appears as singularity over one component. Its Veronese smoothing leads to an additional smoothing component; it exists also for the non hyperelliptic curves.   □

For the case $L = (2g + 2)g_2^1$ one can also use the explicit equations for the base to find the smoothing components. The relevant computations are in [St9, Sect. 3].

Remark. The set $\mathcal{B}$ of subsets T of B modulo the equivalence relation $T \sim B \setminus T$ forms a group, isomorphic to $\mathbb{Z}^{2g+1}$; the subgroup $\mathcal{B}^+$ of T's of even cardinality is isomorphic to $J_2(C)$: every $\eta \in J_2(C)$ can be represented by a divisor $D \sim kg_2^1 - \sum_{i \in T} P_i$, with $|T| = 2k$.

Lemma. *Let $\eta \in J_2(C)$ be represented by $D \sim kg_2^1 - \sum_{i \in T} P_i$, with $T \in \mathcal{B}^+$. Let $S = \mathbb{P}(\pi_* L)$. Then $S_\eta = \mathbb{P}(\pi_*(L \otimes \eta))$ is isomorphic to $\mathrm{elm}_T(S)$.*

Proof. Factorise elm_T as $S \xleftarrow{\sigma} \tilde{S} \xrightarrow{\sigma'} S' = \mathrm{elm}_T(S)$: as before, the exceptional divisors are E_i and E_i'. We have $L = E_0 + af$; let $T_0 \subset T$ be the index set of points P_i on E_0, with cardinality m. Let $\overline{E}_0$ be the strict transform of E_0 on $\tilde{S}$, and E_0' that on S'. Then

$$\sigma^* L = \overline{E}_0 + \sum_{j \in T_0} E_j + af$$

$$= (\sigma')^* E_0' - \sum_{i \notin T_0} E_i' + \sum_{j \in T_0} E_j + af$$

$$\sim ((\sigma')^* E_0' + (a + m - k)f) + \sum_T E_i - kf .$$

On S' the linear system $|E_0' + (a + m - k)f|$ cuts out on C the series $|L + D|$. We have $E_0' \cdot E_0' = -e + 2(k - m)$, so if $2(k - m) < e$, the section E_0' is the unique negative section, and we see directly that $E_0' + (a + m - k)f$ is the hyperplane class ($2a = 3g + 3 + e$). Otherwise there is a section E_- with $E_- \cdot E_- = -e'$, $e' \geq 0$, and $E_0' \sim E_- + (k - m - e/2 + e'/2)f$. The hyperplane class is $E_- + (3g + 3 + e')/2f \sim E_0' + (a + m - k)f$. $\square$

Alternative description of the construction. We have given the surfaces S as ruled surfaces, birationally embedded with the linear system $|C| = |2E_0 + af|$. CASTELNUOVO describes them with linear systems in the plane, of curves of degree $g + e + 1$ with one $(g + e - 1)$-ple point and $e - 1$ infinitely near double points, if $1 \leq e \leq g + 1$, or for $e = 0$ of curves of degree $g + 3$ with a $(g + 1)$-ple point and a double point in a different point [Cas]. If there are at least two finite double points, a standard Cremona transformation will decrease the degree, for details see the book [Con].

Now given a curve C in the plane of degree d with δ double points (finite or not), and with multiplicity $d - 2$ at the origin, and given a set T of Weierstraß points of cardinality k, we form the curve $C \cup (\cup_{i \in T} L_i)$ of degree $d + k$, where L_i is the line joining the origin with the Weierstraß point P_i, and we consider the linear system of plane curves of degree $d + k$ with multiplicity $d + k - 2$ at the origin, and $\delta + k$ double points in the double points of C and in the P_i, $i \in T$. As the curves of this system do not intersect L_i outside C, the rational map $\phi_{|C|}(\mathbb{P}^2) \to \phi_{|C + \sum L_i|}(\mathbb{P}^2)$ is elm_T.

Example. Let X be the cone over the hyperelliptic curve $y^2 - 1 + x^6$, embedded with $6g_2^1$. We describe two surfaces with C as hyperplane section, and identify the induced line in $T_X^1(-1)$. We remark that the Weierstraß points are the sixth roots of unity, and that the group μ_6 operates on $\mathcal{B}$.

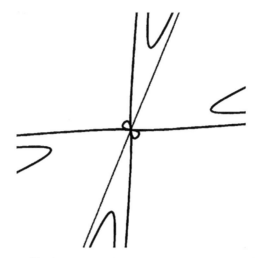

Fig. 16.1. A hyperelliptic curve with a tangent to a Weierstraß point at infinity

We start with the plane curve $x^2y^2z^2 - y^6 + x^6$. The Weierstraß points are the points at infinity. The real picture of this curve is not so interesting. Figure 16.1, made with $surf$[En], shows a different curve of the same type, with a much more complicated equation. It has also a quadruple point at the origin and the Weierstraß points lie on the line at infinity. The linear system containing our original curve consists of curves of degree 6 with multiplicity 4 at the origin O, and A_3 singularities tangent to $x = 0$ and $y = 0$. A basis is

$$z_i = y^{6-i}x^i, \quad i = 0, \ldots, 6 ; \quad w_i = zy^{4-i}x^{i+1}, \quad i = 0, \ldots, 3 ; \quad u = z^2y^2x^2 .$$

This gives equations

$$\text{Rank} \begin{pmatrix} z_0 & \cdots & z_5 & w_0 & w_1 & w_2 \\ z_1 & \cdots & z_6 & w_1 & w_2 & w_3 \end{pmatrix} \le 1 ,$$

$$\phi_m = (w^2)_m - uz_m , \quad m = 0, \ldots, 6 .$$

The image of $\mathbb{P}^2$ is isomorphic to the cone over the rational normal curve of degree 3. With $t = u - z_0 + z_6$ we get the ρ_0 deformation of C, so the ζ_i in the description of the infinitesimal deformations above are zero.

Now consider the Weierstraß point $P_0 = (1 : 1 : 0)$, and the linear system $|C + L_0|$ of septic curves with additional double point at P_0. A basis is given by

$$s_i = (y - x)^2y^{5-i}x^i , \quad i = 0, \ldots, 5 ; \quad u_0 = z^2y^3x^2 , \quad u_1 = z^2y^2x^3 ;$$

$$t_i = (y - x)zy^{4-i}x^{i+1} , \quad i = 0, \ldots, 3 .$$

On C we have $(y - x)^{-1}s_i = z_i - z_{i+1}$, and $(y - x)^{-1}u_0 = z_0 + \ldots + z_5$, $(y - x)^{-1}u_1 = z_1 + \ldots + z_6$; therefore we find as equalities on C

$$6(y - x)z_0 = u_0 + 5s_0 + 4s_1 + 3s_2 + 2s_3 + s_4$$
$$6(y - x)z_1 = u_0 - s_0 + 4s_1 + 3s_2 + 2s_3 + s_4$$
$$6(y - x)z_2 = u_0 - s_0 - 2s_1 + 3s_2 + 2s_3 + s_4$$
$$6(y - x)z_3 = u_0 - s_0 - 2s_1 - 3s_2 + 2s_3 + s_4$$
$$6(y - x)z_4 = u_0 - s_0 - 2s_1 - 3s_2 - 4s_3 + s_4$$
$$6(y - x)z_5 = u_0 - s_0 - 2s_1 - 3s_2 - 4s_3 - 5s_4$$
$$6(y - x)z_6 = u_0 - s_0 - 2s_1 - 3s_2 - 4s_3 - 5s_4 - 6s_5 .$$

With $6(y - x)w_i = 6t_i$ we obtain exactly the same embedding of the curve C as before. To compute the one-parameter we note that C is given by the vanishing of the form $\frac{1}{6}(y-x)^{-1}(u_0 - u_1 - \sum s_i) = \tau$. The matrix for the scroll $\mathrm{elm}_P(S)$ has as first entry $6(y - x)z_0 = u_0 + 5s_0 + 4s_1 + 3s_2 + 2s_3 + s_4$. Below $6(y-x)z_0$ we have to write $u_1 + 5s_1 + 4s_2 + 3s_3 + 2s_4 + s_5 = 6(y-x)z_1 - 6(y-x)\tau$. The second entry on the first row is $u_0 - s_0 + 4s_1 + 3s_2 + 2s_3 + s_4 = 6(y - x)z_1$. In the second part of the matrix we write as entries $6t_i = 6(y - x)w_i$. After dividing by the common factor $6(y - x)$ we obtain a matrix which is a deformation of the original one, and expressed in the deformation parameters ζ_i our one-parameter deformation is given by $\zeta_i = -\tau$ for $i = 1, \ldots, 5$.

The other components are found with similar computations.

Other degrees. The structure of the equations and the dimension of $T_X^2(-2)$, which is $d - 2g - 3$, two less than the number of equations, which cut out the curve on the scroll, lead us to expect that the base space in negative degree is a complete intersection of dimension $4g + 5 - d$, if $2g + 4 \le d \le 4g + 5$. For $d > 4g + 5$ and $L = kg_2^1$ we have a complete intersection of $2g + 2$ quadratic equations [St9, Prop. 3.9], but for other line bunles the number of linearly independent equations will probably be larger.

For $d = 4g + 5$ the cone X has only conical deformations [Te2], and therefore all infinitesimal deformations of negative degree are obstructed. In this case we have indeed a zero-dimensional complete intersection.

For $d = 4g + 3$ the base space should be the cone over a complete intersection curve of degree 2^{2g}, which is a 2^{2g}-fold unbranched covering of C. One finds this degree from the genus formulas. From the point of view of the surfaces, which correspond to one parameter deformations, we get the following picture. A surface S of degree $4g + 3$ is a scroll, blown up in one point P of the hyperelliptic curve C. The fibration $\pi \colon S \to \mathbb{P}^1$ has one singular fibre, which intersects C in P and in $\overline{P}$, the image of P under the hyperelliptic involution. By blowing down one of the two (-1)-curves in the exceptional fibre we get a ruled surface with a minimal section E_0 with $E_0^2 = -e$; if we blow down the other curve, then the parity of e is just the opposite. Suppose that blowing down the curve through P gives an even e. As we have seen, there are 2^{2g} different surfaces with e even, such that the characteristic linear system on C is a given linear system $|L + P|$. Altogether we get a covering of $\mathbb{P}^1$ of degree 2^{2g+1}, with simple branching at the Weierstraß points, so indeed a 2^{2g}-fold unbranched covering of C. In particular, there is only one smoothing component, except of course when $g = 3$.

References

[AC]   Enrico Arbarello e Maurizio Cornalba, *Su una congettura di Petri*. Comment. Math. Helv. **56** (1981), 1–38.

[ACGH]   E. Arbarello, M. Cornalba, P.A. Griffiths and J. Harris. *Geometry of Algebraic Curves*, Vol. I. Berlin etc., Springer 1985.

[AGV]   V.I. Arnol'd, S.M. Gusein-Zade and A.N. Varchenko, *Singularities of Differentiable Maps*, Vol. I. Basel etc., Birkhäuser 1985.

[Al]   A.G. Aleksandrov, *On deformations of one-dimensional singularities with the invariants* $\mathfrak{c} = \delta + 1$. Russian Math. Surveys **33:3** (1978), 139–140.

[Alt1]   Klaus Altmann, *Computation of the vector space* T^1 *for affine toric varieties*. J. Pure Appl. Algebra **95** (1994), 239–259.

[Alt2]   Klaus Altmann, *P-Resolutions of Cyclic Quotients from the Toric Viewpoint*. In: Singularities. The Brieskorn anniversary volume, pp. 241–250. Basel, Birkhäuser 1998 (Progress in Math.; 162).

[Arn]   Jürgen Arndt, *Verselle Deformationen zyklischer Quotientensingularitäten*. Diss. Hamburg 1988.

[Art1]   Michael Artin, *Versal deformations and algebraic stacks*. Invent. math. **27** (1974), 165–189.

[Art2]   Michael Artin, *Algebraic Construction of Brieskorn's Resolutions*. J. Algebra **29** (1974), 330–384.

[Art3]   Michael Artin, *Lectures on Deformations of Singularities*. Bombay, Tata Institute 1976.

[AS]   Enrico Arbarello and Eduardo Sernesi, *Petri's approach to the study of the ideal associated to a special divisor*. Invent. math. **49** (1978), 99–119.

[BC]   Kurt Behnke and Jan Arthur Christophersen, *Hypersurface sections and obstructions (rational surface singularities)*. Compositio Math. **77** (1991), 233–268.

[BE]   David A. Buchsbaum and David Eisenbud, *Some structure theorems for finite free resolutions*. Adv. in Math. **12** (1974), 84–139.

[BG]   Christian Brücker und Gert-Martin Greuel, *Deformationen isolierter Kurvensingularitäten mit eingebetteten Komponenten*. Manuscripta Math. **70** (1990), 93–114.

[Bi1]   Jürgen Bingener, *Offenheit der Versalität in der analytischen Geometrie*. Math. Z. **173** (1980), 241–281.

[Bi2]   Jürgen Bingener, *Lokale Modulräume in der analytischen Geometrie I, II*. Wiesbaden, Vieweg 1987 (Aspekte der Mathematik; D2, D3).

[BKR]   Kurt Behnke, Constantin Kahn and Oswald Riemenschneider, *Infinitesimal Deformations of Quotient Surface Singularities*. In: Singularities, pp. 31–66. Warszawa, PWN 1988 (Banach Center Publications; 20).

[BM]   Dave Bayer and David Mumford, *What can be computed in algebraic geometry?* In: Computational Algebraic Geometry and Commutative algebra, pp. 1–48. Cambridge University Press 1993 (Symp. Math.; 34).

[BR]   Kurt Behnke and Oswald Riemenschneider, *Infinitesimale Deformationen von Diedersingularitäten.* Manuscripta Math. **20** (1977), 377–400, Korrektur ibid. **24** (1978), 81.

[BS]   Dave Bayer and Mike Stillman, *Macaulay: A system for computation in algebraic geometry and commutative algebra.* Computer software, available at http://www.math.columbia.edu/~bayer/Macaulay.

[Bu]   Ragnar O. Buchweitz, Thèse, Paris 1981.

[Buchb] B. Buchberger, *Ein algorithmisches Kriterium für die Lösbarkeit eines algebraischen Gleichungssystems.* Aequationes Math. **4** (1969), 374–383.

[BW]   D.M. Burns, Jr., and Jonathan M. Wahl, *Local contributions to global deformations of surfaces.* Invent. math. **26** (1974), 67–88.

[Cas]   Guido Castelnuovo, *Sulle superficie algebriche le cui sezioni piane sono curve iperellitiche.* Rend. Circ. Mat. Palermo **4** (1890), 73–88.

[Cat]   Fabrizio Catanese, *Commutative algebra methods and equations of regular surfaces.* In: Algebraic Geometry, Bucharest 1982, pp. 68–111. Berlin etc., Springer 1984 (Lect. Notes in Math.; 1056).

[CEP]   Corrado De Concini, David Eisenbud and Claudio Procesi, *Hodge Algebras.* Astérisque **91** (1982).

[CHM]   Ciro Ciliberto, Joe Harris and Rick Miranda, *On the surjectivity of the Wahl map.* Duke Math. J. **57** (1988), 829–858.

[Chr]   Jan Christophersen, *Versal deformations of cyclic quotient singularities.* In: Singularity Theory and its Applications, Warwick 1989, Part I, pp. 81–92. Berlin etc., Springer 1991 (Lect. Notes in Math.; 1462).

[Cob]   Arthur B. Coble, *Point sets and allied Cremona groups* I. Trans. Amer. Math. Soc. **16** (1915), 155–198.

[Com]   Michael Commichau, *Deformation kompakter komplexer Mannigfaltigkeiten.* Math. Ann. **213** (1975), 43–96.

[Con]   Fabio Conforto, *Le Superficie Razionale.* Bologna, Zanichelli 1939.

[Do1]   Adrien Douady, *Déformations régulières.* In: Séminaire Henri Cartan 1960/61.

[Do2]   Adrien Douady, *Obstruction primaire à la déformation.* In: Séminaire Henri Cartan 1960/61.

[Dr]   R. Drewes, *Infinitesimale Deformationen von Kegeln über transkanonisch eingebetteten hyperelliptischen Kurven.* Abh. Math. Sem. Univ. Hamburg **59** (1989), 269–280.

[EGA]   A. Grothendieck, *Éléments de géométrie algébrique.* Rédigés avec la collaboration de J. Dieudonné. IV, *Étude locale des schémas et des morphismes de schémas.* Inst. Hautes Études Sci. Publ. Math. **20** (1964), **24** (1965), **28** (1966), **32** (1967).

[El]   Renée Elkik, *Solutions d'équations à coefficients dans un anneau hensélien.* Ann. sci. École Norm. Sup. (4) **6** (1973), 553–603.

[En]   Stephan Endraß et al. *surf version 1.0.3 — visualizing alegraic curves and algebraic surfaces.* http://surf.sourceforge.net/

[EP]   David Eisenbud and Sorin Popescu, *The projective geometry of the Gale transform.* J. Algebra **230** (2000), 127–173.

[Eph]   Robert Ephraim, *Isosingular loci and the Cartesian product structure of complex analytic singularities*. Trans. Amer. Math. Soc. **241** (1978), 357–371.

[Fi]   Gerd Fischer, *Complex Analytic Geometry*. Berlin etc., Springer 1976 (Lect. Notes in Math.; 538).

[Fl]   Hubert Flenner, *Ein Kriterium für die Offenheit der Versalität*. Math. Z. **178** (1981), 449–473.

[FP]   Jose Ferrer Llop y Fernando Puerta Sales, *Deformaciones de gérmenes analíticos equivariantes*. Collect. Math. **32** (1981), 121–148.

[Gab]   A.M. Gabrielov, *Formal relations between analytic functions*. Functional Anal. Appl. **5** (1971), 318–319.

[Gal]   André Galligo, *Théorème de division et stabilité en géometrie analytique locale*. Ann. Inst. Fourier **29**(2) (1979), 107–184.

[Ge]   Murray Gerstenhaber, *On the deformations of rings and algebras*. Ann. of Math. **79** (1964), 59–101.

[GH]   André Galligo et Christian Houzel, *Module des singularités isolées d'apres Verdier et Grauert*. Astérisque **6–7** (1973), 139–163.

[GL]   V.V. Goryunov and S.K. Lando, *On enumeration of meromorphic functions on the line*. In: The Arnoldfest, Fields Inst. Commun. Vol. **24** (1999), pp. 209–223.

[GM]   William M. Goldman and John J. Millson, *The homotopy invariance of the Kuranishi space*. Illinois. J. Math. **34** (1990), 337–367.

[GPS]   G.-M. Greuel, G. Pfister and H. Schönemann, *Singular 2.0. A computer algebra system for Polynomial Computations*. Centre for Computer Algebra, Kaiserslautern (2001). http://www.singular.uni-kl.de.

[Gra]   Hans Grauert, *Über die Deformationen isolierter Singularitäten analytischer Mengen*. Invent. math. **15** (1972), 171–198.

[Gre]   Mark L. Green, *Koszul cohomology and the geometry of projective varieties*. J. Differential Geom. **19** (1984), 125–171.

[Greu] Gert-Martin Greuel, *On deformations of curves and a formula of Deligne*. In: Algebraic Geometry, La Rábida 1981, pp. 141–168. Berlin etc., Springer 1982 (Lect. Notes in Math.; 961).

[Gro]   Alexander Grothendieck, *Technique de descente et théorèmes d'existence en géometrie algébrique I–VI*. Sém. Bourbaki Nos. 190, 195 (1959/60); 212, 221 (1960/61); 232, 236 (1961/62).

[Ha]   Robin Hartshorne, *Algebraic Geometry*. Berlin etc., Springer 1977.

[Hi1]   Heisuke Hironaka, *Resolution of singularities of an algebraic variety over a ground field of characteristic zero*. Ann. of Math. **79** (1964), 109–326.

[Hi2]   Heisuke Hironaka, *Stratification and flatness*. In: Real and Complex Singularities, Oslo 1976, pp. 199–265. Alphen a/d Rijn, Sijthoff & Noordhoff 1977.

[Jo]   T. de Jong, *Quasi-determinantal rational surface singularities*. Abh. Math. Sem. Univ. Hamburg **69** (1999), 271–281.

[JS1]   Theo de Jong and Duco van Straten, *Deformations of Non-Isolated Singularities*. Preprint, Utrecht 1988. Also in: Theo de Jong, *Non-Isolated Hypersurface Singularities*. Diss. Nijmegen 1988 .

[JS2]   T. de Jong and D. van Straten, *A deformation theory for non-isolated singularities*. Abh. Math. Sem. Univ. Hamburg **60** (1990),177–208.

[JS3]   T. de Jong and D. van Straten, *Deformations of the normalization of hypersurfaces*. Math. Ann. **288** (1990), 527–547.

150   References

[JS4]   T. de Jong and D. van Straten, *On the base space of the semiuniversal deformation of rational quadruple points.* Ann. of Math. **134** (1991), 653–678.

[JS5]   Theo de Jong and Duco van Straten, *A Construction of **Q**-Gorenstein smoothings of index two.* Internat. J. Math. **3** (1992), 341–347.

[Ju]   Michael Junge, *Infinitesimale Deformationen von Quotientensingularitäten (nach zyklischen Gruppen) und deren Obstruktionen.* Diplomarbeit Hamburg 1980.

[Ka]   Sheldon Katz, *The desingularisation of $Hilb^4\mathbf{P}^3$ and its Betti numbers.* In: Zero-dimensional schemes, Ravello 1992, pp. 231-242. Berlin, W. de Gruyter 1994.

[Kl]   Steven L. Kleiman, *The Enumerative Theory of Singularities.* In: Real and Complex Singularities, Oslo 1976, pp. 297–396. Alphen a/d Rijn, Sijthoff & Noordhoff 1977.

[Kle]   Felix Klein, *Vorlesungen über die Entwicklung der Mathematik im 19. Jahrhundert Teil I.* Berlin, Springer 1926 (Grundlehren der mathematischen Wissenschaften; 24).

[Kod]   Kunihiko Kodaira, *Complex Manifolds and Deformations of Complex Structures.* Berlin etc., Springer 1986 (Grundlehren der mathematischen Wissenschaften; 283).

[Kol]   János Kollár, *Flips, Flops, Minimal Models etc.* Surveys in Differential Geometry **1** (1991), 113–199.

[KPR]   H. Kurke, G. Pfister und M. Roczen, *Henselsche Ringe und algebraische Geometrie.* Berlin, VEB Deutscher Verlag der Wissenschaften 1975.

[KS]   K. Kodaira and D.C. Spencer, *On deformations of complex analytic structures* I, II. Ann. of Math. **67** (1958), 328–466.

[KSh]   J. Kollár and N. I. Shepherd-Barron, *Threefolds and deformations of surface singularities.* Invent. math. **91** (1988), 299–338.

[Ku]   Masatake Kuranishi, *Deformations of compact complex manifolds.* Montréal, Les presses de l'univ. de Montréal 1971 (Sém. de Math. Sup.; 39).

[La]   O.A. Laudal, *Matric Massey products and formal moduli I.* In: Algebra, Algebraic Topology and their Interactions, Stockholm 1983, pp. 141-168. Berlin etc., Springer 1985 (Lect. Notes in Math.; 1183).

[Li]   Joseph Lipman, *Double point resolutions of deformations of rational surface singularities.* Compositio Math. **38** (1979), 37–42.

[Lod]   J.-L. Loday, *Cyclic Homology.* Berlin etc., Springer 1992 (Grundlehren der mathematischen Wissenschaften; 301).

[Loo]   Eduard Looijenga, *The smoothing components of a triangle singularity* II. Math. Ann. **269** (1984), 357–387.

[Lop]   Angelo Felice Lopez, *Surjectivity of Gaussian maps on curves in $\mathbb{P}^r$ with general moduli.* J. Alg. Geom. **5** (1996), 609–631.

[MP]   David Mond and Ruud Pellikaan, *Fitting ideals and multiple points of analytic mappings.* In: Algebraic Geometry and Complex Analysis, Berlin etc., Springer 1989 (Lect. Notes in Math.; 1414).

[Mu1]   David Mumford, *Introduction to algebraic geometry. Preliminary version of first 3 Chapters.* s.l., s.a. Published as: *The Red Book of Varieties and Schemes.* Berlin etc., Springer 1989 (Lect. Notes in Math.; 1358).

[Mu2]   David Mumford, *Lectures on Curves on an Algebraic Surface.* Princeton 1966 (Ann. of Math. Studies; 59).

[Mu3]   David Mumford, *A remark on the paper of M. Schlessinger*. Rice Univ. Stud. **59** (1973), 113-117.

[Nij]   Albert Nijenhuis, *On a class of common properties of some different types of algebras* I–II. Nieuw. Arch. Wisk. (3) **17** (1969) I: 17–46, II: 87–108.

[Pa1]   V.P. Palamodov, *Deformations of complex spaces*. Russian Math. Surveys **31**:3 (1976), 129–197.

[Pa2]   V.P. Palamodov, *Cohomology of analytic algebras*. Trans. Mosc. Math. Soc. **2** (1983), 1–61.

[Pa3]   V.P. Palamodov, *Deformations of Complex Spaces*. In: Several Complex Variables IV, pp. 105–194. Berlin etc., Springer 1990 (Encyclopaedia of mathematical sciences; v. 10).

[Pao]   Roberto Paoletti, *Generalized Wahl maps and adjoint line bundles on a general curve*. Pacific J. Math. **168** (1995), 313–334.

[Par]   Guiseppe Parechi, *Gaussian maps and multiplication maps on certain projective varieties*. Compositio Math. **98** (1995), 219–268.

[Pel1]   Ruud Pellikaan, *Finite determinacy of functions with non-isolated singularities*. Proc. London Math. Soc. (3) **57** (1988), 357–382.

[Pel2]   Ruud Pellikaan, *Deformations of hypersurfaces with a one-dimensional singular locus*. J. Pure Appl. Algebra **67** (1990), 49–71.

[Per]   Daniel Perrin. *Courbes passant par m points généraux de* $\mathbf{P}^3$. Mém. Soc. Math. France 28/29, Suppl. au Bull. Soc. Math. France 115 (1987).

[Pi]   Ragni Piene, *Polar classes of singular varieties*. Ann. Sc. Éc. Norm. Sup. **11** (1978) 247–276.

[Pin1]   Henry C. Pinkham, *Deformations of algebraic varieties with* G_m*-action*. Astérisque **20** (1974).

[Pin2]   H. Pinkham, *Deformations of quotient surface singularities*. Proc. Symp. Pure Math. **30**, Part 1 (1977), 65–67.

[Po]   Geneviève Pourcin, *Déformations de singularités isolées*. Astérisque **16** (1974), 161–173.

[PR]   Jürgen Pesselhoy and Oswald Riemenschneider, *Projective resolutions of Hodge algebras: some examples*. Proc. Symp. Pure Math. **40**, Part 2 (1983), 305–317.

[Re1]   Miles Reid, *Canonical 3-folds*. In: Algebraic Geometry, Angers 1979, pp. 273–310. Alphen a/d Rijn, Sijthoff & Noordhoff 1980.

[Re2]   Miles Reid, *Surfaces with* $p_g = 3, K^2 = 4$ *according to E. Horikawa and D. Dicks*. Text of a lecture, Univ. of Utah and Univ. of Tokyo 1989.

[Rie1]   Oswald Riemenschneider, *Deformationen von Quotientensingularitäten (nach zyklischen Gruppen)*. Math. Ann. **209** (1974), 211–248.

[Rie2]   Oswald Riemenschneider, *Familien komplexer Räume mit streng pseudokonvexer spezieller Faser*. Comment. Math. Helv. **51** (1976), 547–565.

[Rie3]   Oswald Riemenschneider, *Zweidimensionale Quotientensingularitäten: Gleichungen und Syzygien*. Arch. Math. **37** (1981), 406–417.

[Rim1]   D.S. Rim, *Formal deformation theory*. In: Groupes de Monodromie en Géometrie Algébrique (SGA 7 I), pp. 32–132. Berlin etc., Springer 1972 (Lect. Notes in Math.; 288).

[Rim2]   Dock S. Rim, *Equivariant G-structure on versal deformations*. Trans. Amer. Math. Soc. **257** (1980), 217–226.

[Rö]   Ancus Röhr, *Formate rationaler Flächensingularitäten*. Diss. Hamburg 1992.

[RV]   Dock Sang Rim and Marie A. Vitulli, *Weierstrass points and monomial curves*. J. Algebra **48** (1977), 454–476.

[Scha]   Mary Schaps, *Versal determinantal deformations*. Pacific J. Math. **107** (1983), 213–221.

[Schl1] Michael Schlessinger, *Functors of Artin rings*. Trans. Amer. Math. Soc. **130** (1968), 208–222.

[Schl2] Michael Schlessinger, *On rigid singularities*. Rice Univ. Stud. **59** (1973), 147-162.

[Schr]   Frank-Olaf Schreyer, *Syzygies of canonical curves and special linear series*. Math. Ann. **275** (1986), 105–137.

[Ser]   Jean-Pierre Serre, *Courbes algébriques et corps de classes*. Paris, Hermann 1959.

[Sev]   Francesco Severi, *Vorlesungen über Algebraische Geometrie*. Leipzig, Berlin, B. G. Teubner 1921.

[St1]   Jan Stevens, *On the number of points determining a canonical curve*. Indag. Math. **51** (1989), 485–494.

[St2]   Jan Stevens, *On the versal deformation of cyclic quotient singularities*. In: Singularity Theory and its Applications, Warwick 1989, Part I, pp. 312–319. Berlin etc., Springer 1991 (Lect. Notes in Math.; 1462).

[St3]   Jan Stevens, *Partial resolutions of rational quadruple points*. Internat. J. Math. **2** (1991), 205–221.

[St4]   Jan Stevens, *Partial resolutions of quotient singularities*. Manuscripta Math. **79** (1993), 7–11.

[St5]   J. Stevens, *The versal deformation of universal curve singularities*. Abh. Math. Sem. Univ. Hamburg **63** (1993), 197-213.

[St6]   Jan Stevens, *Computing versal deformations*. Experiment. Math. **4** (1995), 129–144.

[St7]   Jan Stevens, *Deformations of cones over hyperelliptic curves*. J. Reine Angew. Math. **473** (1996), 87–120.

[St8]   Jan Stevens, *Degenerations of elliptic curves and equations for cusp singularities*. Math. Ann. **311** (1998), 199–222.

[St9]   Jan Stevens, *Rolling factors deformations and extensions of canonical curves*. Documenta Math. **6** (2001), 185–226.

[Str]   Duco van Straten, *Weakly Normal Surface Singularities and Their Improvements*. Diss. Leiden 1987.

[Te1]   Sonny Tendian, *Deformations of Cones and the Gaussian-Wahl Map*. Preprint 1990.

[Te2]   Sonny Tendian, *Surfaces of degree d with sectional genus g in $\mathbb{P}^{d+1-g}$ and deformations of cones*. Duke J. Math. **65** (1992), 157–185.

[Te3]   Sonny Tendian, *Extensions of projective curves*. Amer. J. Math. **116** (1994), 1469–1478.

[Ty]   G.N. Tyurina, *Locally semiuniversal flat deformations of isolated singularities of complex spaces*. Math. USSR-Izv. **3** (1970), 967–999.

[Wa1]   Jonathan M. Wahl, *Simultaneous resolution of rational singularities*. Compositio Math. **38** (1979), 43–54.

[Wa2]   Jonathan Wahl, *Smoothings of normal surface singularities*. Topology **20** (1981), 219–240.

[Wa3]   Jonathan M. Wahl, *The Jacobian algebra of a graded Gorenstein singularity*. Duke Math. J. **55** (1987) 843–871.

[Wa4]   Jonathan Wahl, *Deformations of quasi-homogeneous surface singularities.*
         Math. Ann. **280** (1988), 105–128.
[Wa5]   Jonathan Wahl, *Gaussian maps on algebraic curves.* J. Differential Geom.
         32 (1990), 77–98.
[Wa6]   Jonathan Wahl, *Introduction to Gaussian maps on an algebraic curve.* In:
         Complex Projective Geometry, pp. 304–323. Cambridge University Press
         1992 (London Math. Soc. Lect. Notes Ser.; 179).
[We]   Joachim Wehler, *Versal deformations of Hopf surfaces.* J. Reine Angew.
         Math. **328** (1981), 22–32.

Index

Printing and Binding: Strauss GmbH, Mörlenbach

Vol. 1772: F. Burstall, D. Ferus, K. Leschke, F. Pedit, U. Pinkall, Conformal Geometry of Surfaces in S^4. VII, 89 pages. 2002.

Vol. 1773: Z. Arad, M. Muzychuk, Standard Integral Table Algebras Generated by a Non-real Element of Small Degree. X, 126 pages. 2002.

Vol. 1774: V. Runde, Lectures on Amenability. XIV, 296 pages. 2002.

Vol. 1775: W. H. Meeks, A. Ros, H. Rosenberg, The Global Theory of Minimal Surfaces in Flat Spaces. Martina Franca 1999. Editor: G. P. Pirola. X, 117 pages. 2002.

Vol. 1776: K. Behrend, C. Gomez, V. Tarasov, G. Tian, Quantum Comohology. Cetraro 1997. Editors: P. de Bartolomeis, B. Dubrovin, C. Reina. VIII, 319 pages. 2002.

Vol. 1777: E. García-Río, D. N. Kupeli, R. Vázquez-Lorenzo, Osserman Manifolds in Semi-Riemannian Geometry. XII, 166 pages. 2002.

Vol. 1778: H. Kiechle, Theory of K-Loops. X, 186 pages. 2002.

Vol. 1779: I. Chueshov, Monotone Random Systems. VIII, 234 pages. 2002.

Vol. 1780: J. H. Bruinier, Borcherds Products on O(2,1) and Chern Classes of Heegner Divisors. VIII, 152 pages. 2002.

Vol. 1781: E. Bolthausen, E. Perkins, A. van der Vaart, Lectures on Probability Theory and Statistics. Ecole d' Eté de Probabilités de Saint-Flour XXIX-1999. Editor: P. Bernard. VIII, 466 pages. 2002.

Vol. 1782: C.-H. Chu, A. T.-M. Lau, Harmonic Functions on Groups and Fourier Algebras. VII, 100 pages. 2002.

Vol. 1783: L. Grüne, Asymptotic Behavior of Dynamical and Control Systems under Perturbation and Discretization. IX, 231 pages. 2002.

Vol. 1784: L.H. Eliasson, S. B. Kuksin, S. Marmi, J.-C. Yoccoz, Dynamical Systems and Small Divisors. Cetraro, Italy 1998. Editors: S. Marmi, J.-C. Yoccoz. VIII, 199 pages. 2002.

Vol. 1785: J. Arias de Reyna, Pointwise Convergence of Fourier Series. XVIII, 175 pages. 2002.

Vol. 1786: S. D. Cutkosky, Monomialization of Morphisms from 3-Folds to Surfaces. V, 235 pages. 2002.

Vol. 1787: S. Caenepeel, G. Militaru, S. Zhu, Frobenius and Separable Functors for Generalized Module Categories and Nonlinear Equations. XIV, 354 pages. 2002.

Vol. 1788: A. Vasil'ev, Moduli of Families of Curves for Conformal and Quasiconformal Mappings.IX, 211 pages. 2002.

Vol. 1789: Y. Sommerhäuser, Yetter-Drinfel'd Hopf algebras over groups of prime order. V, 157 pages. 2002.

Vol. 1790: X. Zhan, Matrix Inequalities. VII, 116 pages. 2002.

Vol. 1791: M. Knebusch, D. Zhang, Manis Valuations and Prüfer Extensions I: A new Chapter in Commutative Algebra. VI, 267 pages. 2002.

Vol. 1792: D. D. Ang, R. Gorenflo, V. K. Le, D. D. Trong, Moment Theory and Some Inverse Problems in Potential Theory and Heat Conduction. VIII, 183 pages. 2002.

Vol. 1793: J. Cortés Monforte, Geometric, Control and Numerical Aspects of Nonholonomic Systems. XV, 219 pages. 2002.

Vol. 1794: N. Pytheas Fogg, Substitution in Dynamics, Arithmetics and Combinatorics. Editors: V. Berthé, S. Ferenczi, C. Mauduit, A. Siegel. XVII, 402 pages. 2002.

Vol. 1795: H. Li, Filtered-Graded Transfer in Using Noncommutative Gröbner Bases. IX, 197 pages. 2002.

Vol. 1796: J.M. Melenk, hp-Finite Element Methods for Singular Perturbations. XIV, 318 pages. 2002.

Vol. 1797: B. Schmidt, Characters and Cyclotomic Fields in Finite Geometry. VIII, 100 pages. 2002.

Vol. 1798: W.M. Oliva, Geometric Mechanics. XI, 270 pages. 2002.

Vol. 1799: H. Pajot, Analytic Capacity, Rectifiability, Menger Curvature and the Cauchy Integral. XII,119 pages. 2002.

Vol. 1801: J. Azéma, M. Émery, M. Ledoux, M. Yor, Séminaire de Probabilités XXXVI. VIII, 499 pages. 2003.

Vol. 1802: V. Capasso, E. Merzbach, B.G. Ivanoff, M. Dozzi, R. Dalang, T. Mountford, Topics in Spatial Stochastic Processes. Martina Franca, Italy 2001. Editor: E. Merzbach. VIII, 253 pages. 2003.

Vol. 1803: G. Dolzmann, Variational Methods for Crystalline Microstructure - Analysis and Computation. VIII, 212 pages. 2003.

Vol. 1804: I. Cherednik, Ya. Markov, R. Howe, G. Lusztig, Iwahori-Hecke Algebras and their Representation Theory. Martina Franca, Italy 1999. Editors: V. Baldoni, D. Barbasch. X, 103 pages. 2003.

Vol. 1805: F. Cao, Geometric Curve Evolution and Image Processing. X, 187 pages. 2003.

Vol. 1806: H. Broer, I. Hoveijn. G. Lunther, G. Vegter, Bifurcations in Hamiltonian Systems. Computing Singularities by Gröbner Bases. XIV, 169 pages. 2003.

Vol. 1807: V. D. Milman, G. Schechtman, Geometric Aspects of Functional Analysis. Israel Seminar 2000-2002. VIII, 429 pages. 2003.

Vol. 1808: W. Schindler, Measures with Symmetry Properties.IX, 167 pages. 2003.

Vol. 1809: O. Steinbach, Stability Estimates for Hybrid Coupled Domain Decomposition Methods. VI, 120 pages. 2003.

Vol. 1810: J. Wengenroth, Derived Functors in Functional Analysis. VIII, 134 pages. 2003.

Vol. 1811: J. Stevens, Deformations of Singularities. VII, 157 pages. 2003.

Recent Reprints and New Editions

Vol. 1200: V. D. Milman, G. Schechtman, Asymptotic Theory of Finite Dimensional Normed Spaces. 1986. – Corrected Second Printing. X, 156 pages. 2001.

Vol. 1618: G. Pisier, Similarity Problems and Completely Bounded Maps. 1995 – Second, Expanded Edition VII, 198 pages. 2001.

Vol. 1629: J. D. Moore, Lectures on Seiberg-Witten Invariants. 1997 – Second Edition. VIII, 121 pages. 2001.

Vol. 1638: P. Vanhaecke, Integrable Systems in the realm of Algebraic Geometry. 1996 – Second Edition. X, 256 pages. 2001.

Vol. 1702: J. Ma, J. Yong, Forward-Backward Stochastic Differential Equations and Their Applications. 1999. – Corrected Second Printing. XIII, 270 pages. 2000.

4. Lecture Notes are printed by photo-offset from the master-copy delivered in camera-ready form by the authors. Springer-Verlag provides technical instructions for
the preparation of manuscripts. Macros packages in L^AT_EX2e are available from Springer's web-pages at

http://www.springer.de/math/authors/index.html

Macros in LaTeX2.09 and TeX are available on request at: lnm@springer.de Careful preparation of the manuscripts will help keep production time short and ensure satisfactory appearance of the finished book. After acceptance of the manuscript authors will be asked to prepare the final LaTeX source files (and also the corresponding dvi- or pdf-file) together with the final printout made from these files. The LaTeX source files are essential for producing the full-text online version of the book

(http://link.springer.de/link/service/series/0304/tocs.htm).

The actual production of a Lecture Notes volume takes approximately 12 weeks.

5. Authors receive a total of 50 free copies of their volume, but no royalties. They are entitled to a discount of 33.3 % on the price of Springer books purchased for their personal use, if ordering directly from Springer-Verlag.

6. Commitment to publish is made by letter of intent rather than by signing a formal contract. Springer-Verlag secures the copyright for each volume. Authors are free to reuse material contained in their LNM volumes in later publications: A brief written (or e-mail) request for formal permission is sufficient.

Addresses:
Professor J.-M. Morel, CMLA,
Ecole Normale Supérieure de Cachan,
61 Avenue du Président Wilson, 94235 Cachan Cedex, France
E-mail: Jean-Michel.Morel@cmla.ens-cachan.fr

Professor F. Takens, Mathematisch Instituut,
Rijksuniversiteit Groningen, Postbus 800,
9700 AV Groningen, The Netherlands
E-mail: F.Takens@math.rug.nl

Professor B. Teissier, Université Paris 7
Institut Mathématique de Jussieu, UMR 7586 du CNRS
Equipe "Géométrie et Dynamique", 175 rue du Chevaleret
75013 Paris, France
E-mail: teissier@math.jussieu.fr

Springer-Verlag, Mathematics Editorial, Tiergartenstr. 17,
69121 Heidelberg, Germany,
Tel.: *49 (6221) 487-8410
Fax: *49 (6221) 487-8355
E-mail: lnm@Springer.de